Biochemistry in the Lab

A Manual for Undergraduates

Biochemistry in the Lab

A Manual for Undergraduates

by
Benjamin F. Lasseter

CRC Press is an imprint of the
Taylor & Francis Group, an **informa** business

CRC Press
Taylor & Francis Group
6000 Broken Sound Parkway NW, Suite 300
Boca Raton, FL 33487-2742

Printed on acid-free paper

International Standard Book Number-13: 978-1-138-58992-6 (Hardback)
International Standard Book Number-13: 978-1-138-58996-4 (Paperback)

Library of Congress Cataloging-in-Publication Data

Names: Lasseter, Benjamin F., author.
Title: Biochemistry in the lab : a manual for undergraduates / by Benjamin F. Lasseter.
Description: Boca Raton, Florida : CRC Press, [2020] | Includes bibliographical references and index. | Summary: "Biochemistry in the Lab: A Manual for Undergraduates presents a comprehensive approach to modern biochemistry laboratory teaching together with a complete experimental experience, from molecular biology, cloning, and protein expression, to purification and characterization"-- Provided by publisher.
Identifiers: LCCN 2019026897 (print) | LCCN 2019026898 (ebook) | ISBN 9781138589964 (paperback) | ISBN 9781138589926 (hardback) | ISBN 9780429491269 (ebook)
Subjects: LCSH: Biochemistry--Laboratory manuals.
Classification: LCC QD415.5 .L37 2020 (print) | LCC QD415.5 (ebook) | DDC 572.078--dc23
LC record available at https://lccn.loc.gov/2019026897
LC ebook record available at https://lccn.loc.gov/2019026898

Visit the Taylor & Francis Web site at
http://www.taylorandfrancis.com

and the CRC Press Web site at
http://www.crcpress.com

This book was written for Sumie and the lads.

Here's tae us. Wha's like us? Ōku wa watashitachi no yōde wa arimasen, soshite sorera no subete ga shinde imasu!

Contents

Author Bio

Dr. Benjamin F. Lasseter earned his Ph.D. in biochemistry from Texas A&M University in 2003. He has written several other laboratory manuals for general chemistry students. He presently teaches biochemistry at Christopher Newport University in Virginia, where he manages a research program on biomarkers for oysters. He raises honeybees and tends his forest with his wife and children.

1 Buffers

A large conceptual difference that exists between the way biochemists think and the way organic chemists think has to do with solvents and buffers. The organic chemist will need his reactions to be in an environment of a certain polarity, either a protic environment or an aprotic one, with extreme purity, and at temperatures that range from freezing cold up to hundreds of degrees. Thus, organic chemists agonize over which solvent to use and how to prevent any other chemical contaminating their solvent and how to get a proton to react in a non-polar environment. The biochemist on the other hand, has a much happier and more relaxed lab condition. There is only one solvent for proteins or nucleic acids: water. If you need a source of protons, the 55 M water will always be an excellent source. So far from avoiding contaminating chemicals, the biochemist will deliberately seek them out in order to control the pH of this watery environment. These contaminants, from the organic chemist's point of view, are the buffers the biochemist uses.

A buffer is a chemical system that resists changing pH from a certain value. It is merely a weak acid (HA) and its corresponding weak base (A^-), existing both together in the same environment. The acid component can prevent pH from increasing due to added hydroxide (OH^-) as shown in the following chemical reaction:

$$HA(aq) + OH^-(aq) \rightarrow H_2O(l) + A^-(aq) \quad (1.1)$$

The pH does not increase in this case because there is no OH^- (*aq*) left over. Similarly, the base component can prevent pH from decreasing due to added protons (H_3O^+ in water) as shown in the following chemical reaction:

$$A^-(aq) + H_3O^+(aq) \rightarrow H_2O(l) + HA(aq) \quad (1.2)$$

As before, the pH does not change, this time because there is no H_3O^+ (*aq*) left over. Thus, the pH does not shift very far from a certain value, so long as there are significant quantities of both the acid and base component present. Practically speaking, this resistance to pH change occurs most significantly when the concentration of the acid is no more or no less than ten times the concentration of the base.

The buffer environment is very important to the biochemist because it determines the charge state of the biomolecules. In general, the concentration of protein molecules or nucleic acid molecules will be much less than the concentration of any buffering component. Therefore, they do not cause the pH to become a certain value as much as they become a certain value due to their own pKa values on any ionizable groups. If the pH is less than the pKa, any ionizable side chain will be in its protonated acidic form, with whatever consequences for charge that has. The reverse is true if pH is more than pKa.

Consider: if a protein has some aspartate residues (pKa $\approx$ 4), some lysine residues (pKa $\approx$ 12) and some histidine residues (pKa $\approx$ 6), if it is in an environment where pH = 10, then every one of the aspartate residues will have the carboxylate group deprotonated and with a negative charge, the lysine residues will be protonated with positive charges, and the histidines will all be *deprotonated* with a *neutral* charge. The negatively charged aspartates will be electrically pulled towards the positively charged lysines, but not towards the histidines. Attractions like that are going to affect the shape and function of the protein.

Suppose the same protein were in an environment buffered at pH = 5. The aspartate carboxylates would still be deprotonated and negative, the lysines would still be protonated and positive, but now the histidines would be PROTONATED and now have POSITIVE charge. Now the negatively charged aspartates will be pulled both towards the lysines but also towards the positive charges of the histidines. The structure of the protein would be different at this pH, and the function would change as a consequence. This example demonstrates why the pH value of the buffered environment is important for the protein.

MAKING BUFFERS

Essentially, there are three ways that you can make a buffer. You can add the number of moles of the free acid to give your desired final concentration, and you can titrate with a strong base monitoring the pH until it is the value you want. Similarly, you can add the moles of free base and titrate with strong acid until you are at the pH you want. Either of those methods creates a buffer, because the titration itself creates a solution with both the weak acid and its conjugate weak base. The third method is done by adding the correct number of moles of an acid and its conjugate base to water, knowing that it will automatically adjust the pH to the correct value, and making sure the total buffer concentration is correct.

Method 1: A Free Acid Titrated by NaOH

Suppose you wanted to make 1.5 L of a 0.02 M TRIS buffer (pKa = 8.1) at pH 7.5. You would measure out 0.03 moles of TRIS free acid and put it dry into a beaker. Then you would dissolve the dry TRIS in about a liter of deionized water, and put a pH electrode in to begin monitoring the changes to pH. Next, you would start adding some 0.5 M NaOH, and keep checking the pH as the chemicals mix, until you get a pH of 7.5. Then you add more deionized water until the final volume is 1.5 liters, and your buffer is made.

Method 2: A Free Base Titrated by HCl

Suppose you wanted to make 250 mL of a 0.50 M CAPS buffer (pKa = 10.5) at pH 10.0. You would measure out 0.125 moles of CAPS base and put it dry into a beaker. As before, you would dissolve the dry CAPS in deionized water to approximately 150 mL. Because the buffer is more concentrated than in example 1, you would use a more concentrated reagent to convert the base into the acid form. You would add 2 M HCl until the pH decreases to 10.0, mixing constantly. Then you add deionized water until the final volume is 250 mL. The addition of the acid converted some of the CAPS base form into acid form, and thus you have made your buffer.

Method 3: Adding Free Acid and Free Base to Make the Buffer

This method requires more mathematical calculation in advance than other methods, and is therefore somewhat more difficult for students to do. On the other hand, in practice it results in better buffers, and so should be preferred. Suppose you wanted to make 800 mL of 0.05 M phosphate buffer (pKa = 7.2) at pH 7.8. You have sodium phosphate monobasic (the acid form) and sodium phosphate dibasic (the basic form). This time you have to add both in the right proportion. In order to determine that, you will have to determine the total moles needed using one equation, and then the proportion of moles of acid to moles of base using another equation.

First, the total moles needed are easily calculated by the relationship $n = V * C$.

$$\text{moles phosphate} = 800 \text{ mL} * 0.001 \text{ L/mL}) * 0.05 \text{ M} = 0.040 \text{ moles} \tag{1.3}$$

Additionally, we know that all the moles of phosphate are either the acid or the base, so

$$\text{moles phosphate} = \text{moles acid} + \text{moles base} = 0.040 \text{ moles} \tag{1.4}$$

The Henderson–Hasselbach equation allows us to determine the proportion of acid moles to base moles

$$\begin{aligned} \text{pH} &= \text{pKa} + \log(\text{base/acid}) \\ \log(\text{base/acid}) &= \text{pH} - \text{pKa} \\ (\text{base/acid}) &= 10^{(\text{pH}-\text{pKa})} \\ \text{moles base} &= 10^{(\text{pH}-\text{pKa})} * \text{moles acid} \end{aligned} \tag{1.5}$$

The relationship that was just derived above can be used to substitute for the moles of base in Equation 1.4

$$\begin{aligned} \text{moles phosphate} &= \text{moles acid} + 10^{(\text{pH}-\text{pKa})} * \text{moles acid} \\ \text{moles phosphate} &= \text{moles acid} * (1 + 10^{(\text{pH}-\text{pKa})}) \\ \text{moles acid} &= \text{moles phosphate}/(1 + 10^{(\text{pH}-\text{pKa})}) \end{aligned} \tag{1.6}$$

Equation 1.6 is a general relationship that applies for all conditions of this buffer. Given that we know the moles of phosphate we will use, the pH we want, and the pKa of the buffer, we can plug in all those values.

$$\text{moles acid} = 0.040 \text{ moles}/(1 + 10^{(7.8-7.2)}) = 0.010 \text{ moles acid}$$

Knowing that all the moles of buffer are either moles of acid or moles of base, we now can easily determine the moles of base

$$\text{moles base} = \text{moles phosphate} - \text{moles acid}$$

$$\text{moles base} = 0.040 \text{ moles} - 0.010 \text{ moles} = 0.030 \text{ moles base}$$

At last, the amounts we need to add are clear, and so the procedure is clear as well:

Instructions: Add 0.010 moles of dry sodium phosphate monobasic and 0.030 moles of sodium phosphate dibasic to a beaker. Add deionized water to dissolve the buffer components until the volume is 800 mL. The pH will be 7.8 when it is checked by a pH meter.

Use these methods when you make your buffers. The third method results in the best buffers, but the first two are easier to do, and sometimes you can only obtain either the free acid or free base, not both.

CONCENTRATED BUFFERS

Often, it is not practical to make large volumes of buffer, even if you are going to need large volumes of it later. You might only have 1 liter storage bottles, but need five liters of a buffer. The usual solution is to make a buffer that is more concentrated than you need by some whole-number multiple, and dilute it at the time you need it. The dilution to volume you need is covered by the relationship

$$C_1V_1 = C_2V_2 \tag{1.7}$$

Such buffers are referred to as "X" buffers.

e.g. You need 40 L of 0.001 M TAE buffer that you will use over the course of a week, and you only have a 1 liter storage bottle. Instead of actually making 40 liters, you make one liter of a 0.040 M TAE buffer by one of the methods stored above, and you would call it "40X 0.001 M TAE". Then, every time you need one liter (V_2) at a concentration of 0.001 M (C_2), you take 25 mL (V_1) of your 40X buffer and add 975 mL of deionized water to it, and you have resulted in 1000 mL of a 1X 0.001 M (C_1) TAE buffer.

A List of Common Buffers

Name	K_a	pKa
Phosphate (H_3PO_4)	7.5×10^{-3}	2.1
Phosphate ($H_2PO_4^{-1}$)	6.2×10^{-8}	7.2
Phosphate (HPO_4^{-2})	4.8×10^{-13}	12.3
Carbonic Acid (H_2CO_3)	4.3×10^{-7}	6.4
Bicarbonate (HCO_3^{-1})	4.8×10^{-11}	10.3
acetate	2.5×10^{-5}	4.6
MES	7.9×10^{-7}	6.1
MOPS	7.9×10^{-8}	7.1
HEPES	3.2×10^{-8}	7.5
TRIS	7.9×10^{-9}	8.1
glycine	2.5×10^{-10}	9.6
CAPS	4.0×10^{-11}	10.4

GOAL

In this lab period, you are instructed to prepare

1. 200 mL of 2X 0.125 M CAPS buffer at pH 10.
2. 1 liter of 5X 0.02 M phosphate buffer at pH 7.4.

Unless your instructor says otherwise, then you are to make the CAPS buffer by Method 1 (adding NaOH to a free acid) and the phosphate buffer by Method 3.

CAPS:

a. Obtain a pH meter and calibration solutions. Calibrate the meter following the instructions specific to that model.
b. Determine the mass of CAPS that you must measure out in order to obtain the right number of moles for 200 mL of 2X 0.125 M CAPS. Measure this mass out and place it in a 400 mL beaker.
c. Add 120 mL of deionized water, and dissolve the solid CAPS as much as it will dissolve. You will have to stir for several minutes. It may not dissolve entirely.
d. Obtain some NaOH whose concentration is between 2 M and 5 M. Add approximately 20 mL if the CAPS has not entirely dissolved, and wait for the CAPS to dissolve. If this is not a sufficient quantity for the CAPS to dissolve within a minute of stirring, then add more NaOH.
e. Place the pH electrode into the CAPS solution with constant stirring.
f. Add the NaOH very slowly until the pH is 10.0.
g. Add deionized water until the volume is 200 mL.
h. When you are finished, label your bottle with your initials and exactly which contents are in the bottle and at which concentration. Your bottles will be stored until you use them later in future experiments.

PHOSPHATE:

a. Obtain a 1 L or larger beaker, and a supply of NaH_2PO_4 and Na_2HPO_4. If necessary, any other alkali metal phosphate will also suffice as a replacement.
b. Determine the molar mass of the two kinds of phosphates you have, recognizing that they may be hydrates and will have a different molar mass than an anhydrous mono- or dibasic phosphate. Use this value to calculate the mass of monobasic and dibasic phosphate you will have to add to the beaker.
c. Measure out your calculated masses, and place them in the beaker.
d. Add 800 mL of deionized water. Stir until the phosphate is dissolved.
e. (optional) Use a calibrated pH meter to check the pH of your phosphate buffer. It should be between 6.9 and 7.1 at this point. If not, add 1 M NaOH or 1 M HCl, as needed, until the pH is 7.0.
f. Add deionized water until the volume is 1 L.
g. As with the CAPS buffer, when you are finished, label your bottle with your initials and exactly which contents are in the bottle and at which concentration. Your bottles will be stored until you use them later in future experiments.

OPTIONAL GOAL

You are assigned to test the buffering capacity of the buffers you have made, to determine how much it will resist pH change.
Buffering Capacity: the number of moles of acid or base you can add without having the pH change more than one pH unit.

a. Titrate 20 mL of each of your buffers at 1X concentration with 1 M HCl. Monitor the CAPS buffer using cresol red indicator (which should change from fuchsia to red-orange), and monitor the phosphate buffer using methyl red (which should change from yellow to red).
   - Determine how many moles of HCl were required to reach equivalence for each buffer.
   - Identify which buffer resists more against the lowering of pH.

b. Titrate 20 mL of each of your buffers at 1X concentration with 1 M NaOH. Monitor the CAPS buffer using alizarin Yellow R indicator (which should change from yellow to red), and monitor the phosphate buffer using phenolphthalein (which should change from clear to pink).
   - Determine how many moles of HCl were required to reach equivalence for each buffer.
   - Identify which buffer resists more against the raising and the lowering of pH.
   - Identify each buffer's capacity against addition of protons or addition of hydroxide.

Suggested materials for set up: CAPS free acid or free base; sodium phosphate monobasic; sodium phosphate dibasic; 1 M HCl; 1 M NaOH; 3 M NaOH; deionized water; pH meter; pH meter standards; stirring rods.

PRE-LAB QUESTIONS

1. You are assigned to make 500 mL of a 0.075 M phosphate buffer, pH 7.5. You have available solid anhydrous potassium phosphate monobasic, KH_2PO_4, and solid sodium phosphate dibasic heptahydrate $Na_2HPO_4 \cdot 7\,H_2O$. Determine the mass of each component you will have to add in order to make your buffer as assigned.

2. Identify by pKa buffer materials that would be good for making buffers of the pH values shown in the following:

Desired pH	Buffer Component	pKa
3.0		
4.0		
5.0		
6.0		
7.0 (*e.g.*)	NaH_2PO_4	7.21
8.0 (*e.g.*)	TRIS	8.06
9.0		
10.0		
11.0		
12.0		

POST-LAB QUESTIONS

1. The reaction of isopropylmalate dehydrogenase is as follows:

   isopropylmalate + NAD^+ $\rightarrow$ 2-oxoisocaproate + CO_2 + NADH + H^+

   CAPS buffer pH 10 works better for this reaction, due to something specific to the dehydrogenation, which can be seen in the reaction itself. If you use CAPS buffer at pH 10, then the initial rate of the reaction can be studied without difficulty. If you use phosphate buffer pH 7, the reaction shows significant interference at all concentrations. If you use glycine buffer with a pH around 4, then the rate of the reaction is so slow as to make observation almost impossible. Furthermore, if you use 1 M CAPS buffer, then the initial rate of the reaction can be studied without difficulty, but if you use 0.001 mM CAPS buffer, then there is significant interference to the initial rate shortly after the reaction is initialized, even though both buffers are initially at pH 10. Explain why.

2. Phosphate buffer, pH 7, is often a good choice for buffering tissue initially in the purification of proteins, especially muscle or heart tissue. Why would it be a better choice than acetate buffer at pH 4.5 or CAPS buffer at pH 10?

3. Different indicators were recommended for use in testing the buffering capacity. Why were these specific indicators recommended?

2 Assays

Assay: A test to reveal the presence of something, and its quantity.

If you are going to do a protein purification, it is of the first importance that you have some kind of assay for your protein, or you will never find it. The different kinds of assays that exist have an extraordinary range of types. There are assays by color, assays by the activity of an enzyme, activity by binding to some ligand, assays by immunochemistry, and far more. Before ever attempting a purification of a protein, work out how you are going to find it.

There will be considerations you must take into account in developing your assay. One is how specifically you need to determine your protein from any other possible isozyme. Another is the cost and availability of the resources you must obtain in order to do the assay. The third is the technical difficulty and reproducibility of the assay. There are other considerations, but you will determine what they are as you do them.

Several types of assays are described, so that you can see the principles of design involved.

1. LACTATE DEHYDROGENASE: ASSAY BY SUBSTRATES ALONE

There are two versions of the enzyme Lactate Dehydrogenase (LDH), L-LDH and D-LDH. They are so named because they oxidize, respectively, the L- and D- isomers of lactate to pyruvate, and vice versa. In both cases, the reaction can be written as:

$$\text{Pyruvate} + \text{NADH} + \text{H}^+ \rightleftarrows \text{Lactate} + \text{NAD}^+ \tag{2.1}$$

NADH is widely used for spectroscopic measurements of enzymes. There is a great difference between the absorbance at 340 nm for NADH, which absorbs strongly, and NAD^+, which does not. In the reaction above, the moles of reaction can be determined by measuring the gain or loss of absorbance at 340 nm. Because of this principle, the assay for LDH is quite simple. A *cocktail* of pyruvate and NADH in a low pH

> Cocktail – a nickname for some combination of chemicals in solution, generally applied to solutions used for assays. The name comes from the glassware used to develop the first assays for enzymes: only cocktail glasses were available, so they were used. The nickname has remained ever since.

buffer is prepared and put into a spectroscopic cuvette. A sample, possibly containing LDH, is added, and the absorbance is measured. If the absorbance decreases, then it indicates the LDH is present in the sample, since no other enzyme would oxidize NADH with only pyruvate present. The assay designed this way cannot discriminate between the two enzymes L-LDH and D-LDH. The enzymes are completely unrelated to each other, and usually purify differently, but if there is no reason to care which enzyme you are getting, or if the only materials available are pyruvate and NADH, then this is an excellent assay.

To be more specific, a cocktail could be made of L-lactate and NAD^+ in a high pH buffer. In this case, you would monitor for the gain of absorbance at 340 nm. When the sample is added, if L-LDH is present, then there would be a gain of absorbance, but if R-LDH or any other enzyme is present, there would be none. Without the right substrate, there cannot be anything to help reduce the NAD^+. This kind of assay is much more specific, and so would be a better-designed one, if all else is equal. A similar assay could be made specifically for D-LDH, if D-lactate is used instead of L-lactate.

This assay, being spectrophotometric, relies upon Beer's Law of absorbance

$Abs = \varepsilon\lambda c$, where c is the concentration of NADH, λ is the path length and is 1.00 cm in most cuvettes, and ε is the intrinsic absorptivity of NADH, which is 6220 $M^{-1}\,cm^{-1}$ at 340 nm.

Additionally, it requires one to know the volume of the assay cocktail, which is variable by the experimenter. Since the activity of this enzyme depends upon the formation of NADH, the change in absorbance will correspond to a change in concentration, which can be expressed as

$$\Delta c = \Delta Abs/\varepsilon\lambda$$

Multiplied by the volume of the activity cocktail and with some unit correction, this change of concentration will reveal the activity of the enzyme.

Activity – the number of micromoles per minute of reaction catalyzed. Sometimes this is confused with the micromoles per minute of substrate reacted, and sometimes it is confused with the rate of product formed. In fact, it is usually numerically equal to one or the other, but the more accurate understanding is the number of reactions itself, not just the amount of substrate reacted or product formed. *e.g.* The reaction 2 ADP → ATP + AMP has a stoichiometry of 2 ADP per reaction. If 26 μmole of ADP are reacted per minute, the activity is half that value, or 13 μmole/minute. This is also just called "13 units of activity" or "13 U" by most biochemists. The observed activity is also modified if the enzyme has a dilution factor (see a description of "dilution factors" in the chapter discussing buffers. For example, if the enzyme above showed 13 units of activity, but it was diluted with a dilution factor of 500, then the actual activity is 6500 U (*i.e.* 13 U × 500), not 13 U. Be certain to make any dilutions with the assay buffer and NOT with water or the activity cocktail itself!

2. PHOSPHOFRUCTOKINASE: ASSAY BY METABOLIC PATHWAY

Phosphofructokinase-1 (PFK) is an enzyme that is part of the glycolysis pathway. It catalyzes the reaction converting Fructose-6-phosphate (F6P) and

Adenosine-5′-triphosphate (ATP) into Fructose-1,6-bisphosphate (F16BP) and Adenosine-5′-diphosphate (ADP). The reaction is shown below:

$$\text{F6P} + \text{ATP} \rightleftarrows \text{F16BP} + \text{ADP}$$

The reaction also has a magnesium requirement. However, none of the components are optically active at any easily measured wavelength. Because of this, any assay is going to have to be somewhat more complex. In this case, it is possible to connect the products of PFK to other more easily measured components by means of enzymes in a metabolic pathway. F16BP can be converted rapidly to glyceraldehyde-3-phosphate (G3P) and dihydroxyacetone phosphate (DHAP) by the enzyme aldolase. G3P and DHAP can be readily interconverted by means of the enzyme triose phosphate isomerase (TIM). DHAP can then be reduced in the presence of NADH to glycerol-3-phosphate and NAD^+ by means of the enzyme glycerol-3-phosphate dehydrogenase (Gl3PDH). This pathway is shown in Figure 2.1.

Thus, a cocktail which is buffered with F6P, ATP, some magnesium, NADH, and abundant quantities of aldolase, TIM, and Gl3PDH would be ideal for identifying the presence of PFK. If the cocktail is put into a spectroscopic cuvette, and a sample possibly containing PFK is added, then the loss of absorbance at 340 nm would be indicative of PFK. No other way exists for F16BP to be produced, and every mole of it will quickly be converted to G3P and DHAP. The DHAP will quickly be reduced, and the G3P will also be reduced as it is converted to DHAP, too. Whenever the DHAP is reduced, the NADH is converted to NAD^+, resulting in a loss of absorbance. There is a 2 to 1 stoichiometric proportion of NADH oxidized to F6P converted, and that is factored into the calculation of how much PFK is present. For every two micromoles of NADH oxidized per minute, there is one unit of enzyme activity.

Students who have studied molecular biology will generally be familiar with "selection", "knockout mutations", or "survival recovery". Each of these molecular biology tools is actually a variation of an assay by metabolic pathways in which a vital function is gained or lost, and detected by whether the organism lives or dies.

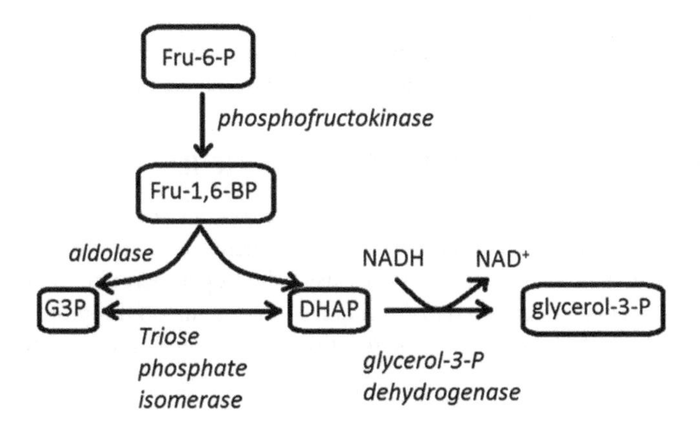

FIGURE 2.1 The use of enzymes in an established metabolic pathway allows the consumption of NADH to be coupled to the activity of phosphofructokinase.

3. ASSAY BY IMMUNOCHEMISTRY

Antibodies are proteins that specifically bind certain chemicals. They can be produced monoclonally against any protein, whether it has enzymatic activity or not. This is very useful when you are trying to purify a protein which has no activity to test for. Such is often the case if you are after perhaps a signaling protein, which would have no effect on its own, and you want to purify it separately from any other protein in a signaling pathway to which your protein of interest might bind. If there is any way to visualize the antibody, it can be used to locate your protein. A number of ways exist to visualize the antibody. If they are grown in the presence of radioactive sulfur, they will have inherent radioactivity and can be detected that way. That is not often preferred by modern biochemists, both due to the cost of radioactive materials and because of safety concerns, but it does work. It is also possible to connect the gene for the antibody to the gene for some easily visualized enzyme such as horseradish peroxidase, which will create a bright color change in the presence of certain substrates. This method also works well, but the substrate has an added cost.

Green fluorescence protein (GFP) is a bioluminescent protein first isolated from the jellyfish *Aequora victoria*. When ultraviolet light illuminates GFP, it glows a bright green color. GFP can be linked to any antibody the same way the horseradish peroxidase can – at the genetic level. Wherever the antibody goes will glow green upon exposure to ultraviolet light. The antibody can be specific for your protein.

This sort of assay will be somewhat different because it does not use a solution cocktail. The protein can be separated onto some semisolid surface such as a polyacrylamide gel ("PAGE", which you can read about in a later chapter). Then the surface itself is treated with a solution containing the GFP-linked antibody. A light washing will remove all antibody that is not bound to your protein. Then, the surface can be illuminated with ultraviolet light. Any position that glows green indicates the presence of your protein. The more intense the green color is, the higher the concentration of your protein. Other reporter systems include chromogenic enzymes such as alkaline phosphatase and horseradish peroxidases, or even radiolabelled antibodies detected by radioimaging.

This method can be very costly, but it is extremely useful for proteins that are otherwise hard to find. A version of it is described in the next chapter.

4. CATALASE: ASSAY BY PHYSICAL PROPERTIES

Some enzymes can be assayed without any spectroscopic equipment at all. They can be detected using physical properties that are themselves quite easy to perform. Catalase is one such enzyme. Catalase converts hydrogen peroxide into water and oxygen. It is part of a system used to eliminate oxidizing materials before they can do cell damage in aerobic organisms. The reaction it catalyzes is given below:

$$2H_2O_2 \rightarrow 2H_2O + O_2(g)$$

Because it produces a micromole of oxygen gas per minute for every ***unit*** of activity, the activity can be measured directly by measuring the volume of gas produced, or indirectly by measuring how quickly that gas can change the density of anything it clings to. Either method is good, though the direct measurement of volume is more accurate and the measurement of density change is less expensive and easier to prepare.

In either case, the moles of oxygen equal the moles of reaction, and they may be determined by the ideal gas law:

$$n = PV/RT$$

where R is the universal gas constant 0.08206 L·atm/mol·K, P is the pressure, V is the volume of the gas, and T is the temperature.

In general, if an enzyme can be measured by something simple instead of something complicated, the simpler approach is to be preferred. Other simple assays involve monitoring pH change, with indicators or with pH meters. Assays such as this will usually have to be done in unbuffered conditions, which is a drawback: enzymes will not remain stable long in such circumstances, so the work must be done quite rapidly.

Ways of Reporting Enzyme Activity

There are several similar sounding terms that pertain to enzyme activity. Each one is used for a slightly different purpose:

Activity: The micromoles per minute of enzymatic reaction catalyzed; the Units (U) for the enzyme. This reports on whether the enzyme is present or not. It also is an essential part of all other ways of knowing the activity. What has been described up to this point has just been "activity" and not any of the other terms.

Relative Activity: the concentration of units of activity, expressed as units/mL (U/mL). This is a direct measure of the quantity of *active* enzyme in a solution. In combination with knowing the total volume of the enzyme source solution, it allows the experimenter to determine the amount of active enzyme that can be purified, in total.

Total Activity: The total number of units of active enzyme in a sample. Though it is reported in Units (U), just like the activity, it represents something slightly different. The activity can indicate something about the presence of the enzyme, but the total activity represents the total number of copies of that enzyme that can be obtained out of a sample. Under optimal circumstances, it will never diminish from the beginning of any enzyme purification until the end.

Specific Activity: The micromoles per minute of enzymatic reaction catalyzed, divided by the mass of enzyme used; it is a measure of the fraction of proteins that are the enzyme for which you are assaying, or perhaps the purity of your enzyme within the sample. It is reported in "units per milligram" (U/mg). The specific activity will not be measured in this lab, because it requires you to know the concentration of your protein in addition to the activity and relative activity.

Evaluating a Source for an Enzyme

The total activity is determined by multiplying the relative activity by the total volume. Determine the total activity for each enzyme from each source. Then evaluate which source is better for obtaining the enzyme, based upon the total activity and the mass of the source used.

e.g. Group A received 12.0 mL of extract from 5.22 g of beef and relative activity for L-LDH of 2450 U/mL and relative activity for catalase of 78000 U/mL. Group B received 10.5 mL of extract from 4.81 g of fish, but measured relative activities of 2440 U/mL and 68000 U/mL for L-LDH and catalase, respectively. They would have total activities (TA) as shown in below:

$$\text{Group A L-LDH: TA} = 2450\ \text{U/mL} * 12.0\ \text{mL} = 29400\ \text{U}$$

$$\text{Group B L-LDH: TA} = 2440\ \text{U/mL} * 10.5\ \text{mL} = 25620\ \text{U}$$

$$\text{Group A catalase} = 78000\ \text{U/mL} * 12.0\ \text{mL} = 936000\ \text{U}$$

$$\text{Group B catalase} = 68000\ \text{U/mL} * 10.5\ \text{mL} = 714000\ \text{U}$$

In terms of evaluating how good a source each one was, it would then be appropriate to divide the total units of activity by the mass from which the enzyme came.

$$\text{Group A L-LDH: TA} = 29400\ \text{U}/5.22\ \text{g} = 5630\ \text{U per gram}$$

$$\text{Group B L-LDH: TA} = 25620\ \text{U}/4.81\ \text{g} = 5330\ \text{U per gram}$$

$$\text{Group A catalase} = 936000\ \text{U}/5.22\ \text{g} = 179000\ \text{U per gram}$$

$$\text{Group B catalase} = 714000\ \text{U}/4.81\ \text{g} = 148000\ \text{U per gram}$$

Then an evaluation could be made as to which source is the best. In this case, both Group A's beef and Group B's fish are approximately equal as sources for L-LDH, though there is a very slight preference for the beef in this case. On the other hand, there is almost a fifth more catalase in the beef than in the fifth, making it the best source, if all other factors are equal. At this point the researcher would also consider factors such as the relative cost of beef versus fish, whether there are legal or ethical restrictions in using one or the other, or any other factor beyond the mere consideration of quantity of protein from a given amount of material.

Serial Dilutions

If you need to dilute anything by a factor of 50,000, it is not always practical to do so in one step. If you are making a final volume of only 1 mL for a solution of some glucose, then it is not possible to accurately pipet a volume of 0.00002 mL of glucose into 0.99998 mL of buffer with any accuracy. Instead, you would do it in a series of steps, each step being quite accurate:

Pipet 0.1 mL of glucose into 0.9 mL of buffer to make "solution 1".
Pipet 0.1 mL of solution 1 into 0.9 mL of buffer to make "solution 2".
Pipet 0.1 mL of solution 2 into 0.9 mL of buffer to make "solution 3".
Pipet 0.1 mL of solution 3 into 0.9 mL of buffer to make "solution 4".

Then, pipet 0.2 mL of solution 1 into 0.5 mL of buffer to make your final solution. Each of the first four solutions is diluting by a factor of ten, such that solution 1 is diluted by 10, solution 2 is diluted by 100, solution 3 is diluted by 1,000, and solution 4 is diluted by 10,000. The last solution dilutes solution 4 by a factor of five, for a final dilution of 50,000. The volumes are determined by the classic equation from general chemistry

$$C_1V_1 = C_2V_2$$

An image representing this serial dilution is shown in Figure 2.2.

Goal

Assay an extract of calf liver, an extract of fish muscle, and an extract of carrot root for **L-lactate dehydrogenase** and for **catalase**. Determine the relative activity and total activity of each enzyme in each extract. Determine which is the best source to use for each enzyme.

Materials

5 g of liver
5 g of fish muscle
5 g of carrots
250 mL of 0.02 M phosphate buffer, pH 7.0
A commercial blender
A preparative centrifuge with a JA-20 or JA-14 rotor or equivalent

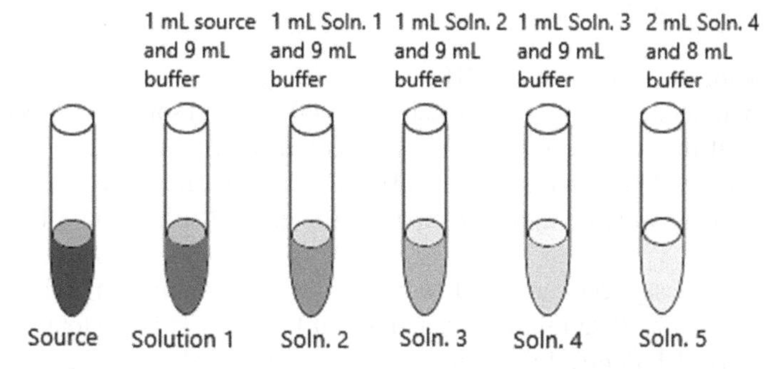

FIGURE 2.2 Serial dilution lowers the concentration of the source more accurately than single-step dilution.

Centrifuge tubes for a JA-20, adapters if using a JA-14

Instructor: for each protein source place 5 g in 35 mL of buffer per student group, and grind in the blender for up to two minutes. Balance the masses of the tubes and centrifuge at 18,000 × g for 15 to 20 minutes. Save the supernatant on ice, and distribute at least 30 mL samples to the student groups.

A UV/VIS spectrophotometer per group
Two or more matched cuvettes per group
6 mM NAD^+
150 mM L-lactate
0.125 M CAPS buffer pH 10.0
P1000, P200, and P10 (or P20) adjustable pipets, with appropriate tips
A beaker for the cocktail

Instructor: turn on the spectrophotometer before the lab begins so that the lamp is warm. Adjust the wavelength to 340 nm for absorbance studies.

Commercial 3% hydrogen peroxide
Paper disks stamped by a hole puncher from Whatman40 filter paper
Forceps or tweezers
10 mL graduated cylinder
Stopwatch
50 mL beakers

INSTRUCTIONS

Before beginning any step, record the total volume of the enzyme sample that has been dispensed to you. Ask your instructor to tell you the mass of the source material and the volume he or she used to make the initial extraction. When you do your own purifications later, these are quantities you must always record. In this lab session, you are determining the activity at any concentration using a single measurement. In later sessions, you will find that you obtain more accurate results when you measure the activity in triplicate.

1. Assay for the activity of L-LDH
   You will need a spectrophotometer set to 340 nm, and some cuvettes transparent at that wavelength. Your solutions will be 6 mM NAD^+, 150 mM L-lactate, 0.125 M CAPS buffer at pH 10.0. The experiment is easiest if you also have a 1000 μL adjustable pipet as well as a 100 μL or 200 μL adjustable pipet.
   For each assay cocktail, you would normally combine in the cuvette 0.95 mL buffer, 0.25 mL lactate, and 0.25 mL NAD^+. This is a total volume per assay of 1.45 mL. In practice, however, you should determine the number of assays you intend to do first, multiply the volumes listed by 110% of the number of samples you intend to do, so that you have a slight amount

left over. *e.g.* You know you need to do 20 samples. Therefore, you combine 20.9 mL of buffer (*i.e.* 0.95 mL × 20 samples × 110%), 5.5 mL of lactate, and 5.5 mL of NAD^+. You mix this and store it on ice until you are ready to use it, and every time you are ready to do an assay, you withdraw 1.45 mL from the cocktail you have prepared and add it to the cuvette.
When you are ready to perform the assay and have 1.45 mL of cocktail in the cuvette, blank the cuvette at 340 nm, and then add 0.050 mL of your enzyme at whatever dilution you are using. (Note that your reaction volume is now 1.50 mL, or 0.00150 L.) Immediately mix the enzyme and begin monitoring the absorbance as it increases over the next minute. The activity can be determined as described above. *e.g.* suppose that in one minute there was an increase in absorbance of 0.203 units with an enzyme whose dilution factor was 500. The activity would be:

$$U = 0.203/(\text{min} * 6220\ M^{-1}cm^{-1} * 1.00\ cm)$$

$$\times (10^6\ mM/M) * 0.00150\ L * 500 = 24.5\ U$$

All other calculations in this assay are similar. The only parts that change from experiment to experiment is the change of absorption and the dilution factor. The rest is constant if you are using the same equipment, materials, and volume.

2. Assay for the activity of catalase
You will need some 50 mL beakers, a 10 mL graduated cylinder or a 10 mL transfer pipet, a stock of 3% hydrogen peroxide, a stopwatch, some forceps or tweezers, a number of paper disks of Whatman 40 filter paper punched out by a standard notebook hole puncher, and a 10 μL adjustable pipettor. Determine the room temperature and the pressure when lab begins, and provide this information to the students. Into the clean beaker, add 10 mL of hydrogen peroxide. Pick up a paper disk with the forceps. With the pipettor, add 5.00 μL of extract to the paper. Drop the paper into the hydrogen peroxide, and start the stopwatch right away. The paper should sink immediately, and bubbles of oxygen will start to form and cling to the paper. When the density of the paper is less than that of water, it will float to the surface. Stop the time at that point. Record the temperature and pressure.

In practice, if it takes between 10 seconds and 2 minutes for the paper disk to float, the data are more reliable. If the disk rose in less than 10 seconds, then diluting the enzyme would give better results. If the disk took longer than 2 minutes to rise, there probably is no enzyme present, or else an enzyme too dilute to be of use.

To a crude estimation, the volume of oxygen gas bubble that must cling to the paper in order to make it float is 10 μL. Knowing the temperature and pressure allows you to calculate the micromoles of oxygen gas produced, which is in a 1 to 1 proportion to the micromoles of reaction catalyzed. The combination of these data allows us to calculate the activity of the enzyme. *e.g.* Suppose the extract had a dilution factor of 500, and it took 32 seconds for the paper disk to rise to the surface, when the temperature was 19°C

(292 K) and the barometric pressure was 0.997 atm. The activity would be determined as follows:

$$U = [(0.997 \text{ atm} * 10 \times 10^{-6} \text{ L})/(0.08206 \text{ L} \cdot \text{atm/mol} \cdot \text{K} * 292 \text{ K})]$$

$$\times (10^{6} \text{ mmol/mol}) * (60 \text{ s/min} * 1/32 \text{ s})$$

$$= 0.780 \text{ U} * 500$$

$$= 390 \text{ U}$$

All calculations for this assay are done similarly. You will probably want to prepare a number of beakers with hydrogen peroxide in advance, and also enzyme dilutions with dilution factor of 10, 50, 100, 500, and 1000. Make these dilutions with the same buffer the extraction was done in, generally a phosphate buffer around pH 7.

***Alternative test (Volume Displacement): Obtain a 20 mL Erlenmeyer flask with a rubber stopper. Insert a #2-gauge needle from smaller face to the larger one, so that it pierces through the stopper. Cut 10 cm of silicone tubing, and place it over the needle at one end of the tubing. Attach the other end of the tubing to a low-resistance 1 mL syringe. Place 9.990 mL of 3% hydrogen peroxide in the flask. Add 10 μL of your dilution of catalase to the hydrogen peroxide. Quickly swirl the flask and cap it. Measure the change of volume in the syringe for two minutes, being sure to record the exact length of time that elapses. Check the room temperature and pressure. The volume change in the syringe will be the volume of oxygen produced, and the temperature and pressure of the room will be the temperature and pressure of the oxygen. Note that any resistance from the syringe will actually add to the pressure of the oxygen, so be sure to use a syringe that offers a minimum of resistance, or you will need to find some means to correct the pressure.*

3. Find the relative activity of each enzyme from each source

L-LDH: Many first-time biochemists get confused when determining the activity in an assay cocktail. There are multiple volumes that all seem important – the 1.50 mL of the cocktail and the 10 μL of the enzyme dilution that were used. This is a time when the distinction between the *activity* and the *relative activity* will now become apparent. When determining the activity of L-LDH, the important volume was the volume of the cocktail, 1.50 mL. However, in finding the relative activity, the important volume is now the volume of the enzyme dilution used, 10 μL. The relative activity is the activity divided by the volume used.

e.g. the activity found using 10 μL of the enzyme dilution was 24.5 U, after accounting for the dilution factor. The relative activity (RA) would be:

$$RA = (24.5 \text{ U}/10 \text{ μL}) * (1000 \text{ μL/mL}) = 2450 \text{ U/mL}$$

Catalase: This is extremely similar to finding the relative activity of L-LDH, except you will have used a different volume of enzyme if you use the paper disk method (5 μL) than if you use the alternative volume displacement test. Be certain to have recorded the actual volume you use each time.

e.g. The activity found using the Paper Disk Method was 390 U and 5 μL were used to spot the paper. The relative activity (RA) would be:

$$RA = (390\ U/5\ \mu L) * (1000\ \mu L/mL) = 78000\ U/mL$$

4. Find the total activity
   Determine the total activity of each enzyme in each source. Then evaluate which source is the best one to choose in order to begin purification of an enzyme, all other factors beyond those studied in this lab being considered equal.

A table such as the following is a good idea to keep in your notebook

Source	Tot. Volume	Dil. for L-LDH	L-LDH Activity	L-LDH Rel. Act	L-LDH Tot. Act	Dil. for Catalase	Catalase Activity	Catalase Rel. Act.	Catalase Tot. Act.

In later laboratory periods you will also include columns for protein concentration and "specific" activity.

PRE-LAB QUESTIONS

1. You extract 4.735 g of tissue from an alligator heart in 100 mL of buffer and take a 10.00 mL aliquot of it for your own use. You test it for L-LDH activity using the technique described in this lab book, using a 1:100 dilution. The Abs_{340} increases from 0.0016 to 0.0676 over 60 seconds.
   a. Determine the activity.

   b. Determine the relative activity.

   c. Determine the total activity.

2. Explain why the relative activity depends upon the volume of the enzyme used and not the volume of the enzyme assay cocktail.

POST-LAB QUESTIONS

1. Find the reaction that is catalyzed by isopropylmalate dehydrogenase. Identify the components that should be in a reaction cocktail if you are assaying it by activity. You should be able to identify appropriate buffer conditions as well as every component necessary to detect the presence of the enzyme, though you may ignore concentrations of the components for this question.

2. Find the reaction that is catalyzed by pyruvate kinase. This reaction cannot be directly visualized by spectroscopy, because there is no optically active component involved in the reaction. However, by the addition of NADH, MgADP, phospho(enol)pyruvate, and *abundant* L-LDH, it can be easily detected, using the linked metabolic pathway.
   a. Sketch the reaction pathway that will be involved in this cocktail.

   b. Why does the measurement of the L-LDH reaction actually reveal the pyruvate kinase reaction in this cocktail?

   c. Why must the L-LDH be ***in great abundance*** for this assay to work?

3. Out of beef liver, fish muscle, and carrot root, one of these sources is the least good for extracting catalase and one was the least good for extracting L-LDH. Determine the biological function of each enzyme and explain why the source would innately be a poor source for each enzyme.

3 Protein Concentration

There are many reasons why you would need to calculate the concentration of a protein sample. When you are doing protein purifications, you need to know the concentration to evaluate how pure your sample is. When you crystallize proteins, you need to know the concentration. Most molecular biology techniques using proteins require very specific concentrations, which again you need to determine. When you do enzymology, you must know the concentration of the protein to calculate some of the parameters. As said, there are many reasons why you need to do this.

A wide range of techniques exist to determine protein concentration, spectroscopically. Each one has some advantages and disadvantages compared to the others. All of them are based upon Beer's Law, which states:

$$\text{Abs} = \varepsilon \lambda c \tag{3.1}$$

where c is the concentration, λ is the path length of light in the spectrophotometer, and ε is the extinction coefficient for your protein at that wavelength. Conceptually, an extinction coefficient is the intrinsic amount of light at a particular wavelength that a substance is capable of absorbing. Rubies have a very low value of the extinction coefficient for red light and a very high one for blue light, as an example.

Some of the methods to determine protein concentration are described in the following sections.

1. ABSORBANCE AT 280 NM

Residues of tryptophan, tyrosine, and phenylalanine all absorb ultraviolet light at 280 nm somewhat well, as do cysteine residues formed from a pair of cysteines. Their extinction constants are given in Table 3.1.

If a protein had one tryptophan, five tyrosines, six phenylalanines, and two cysteine disulfide bonds, then that protein would have an extinction coefficient of 13206.

In a highly pure sample of a protein, this can be calculated by multiplying each residue's extinction coefficient by the number of such residues, and then merely adding them together. The protein can be quantified then by putting it into a quartz cuvette and measuring the absorbance at 280 nm, using light produced in a UV/VIS spectrophotometer by a deuterium lamp. Generally, protein samples are not pure at all, and to determine the protein concentration in mg/mL, a standard curve is made with a known protein such as albumin. It is assumed that on average all proteins would have the same proportion of absorbing residues relative to their mass, and so a milligram of any protein would absorb the same as a milligram of albumin.

This method is extremely accurate under the right circumstances. However, it has some drawbacks which make it not always the method of choice.

First of all, the listed extinction coefficients are not always correct. If the environment around a residue is abnormally polar or non-polar, or if the residue is especially mobile, or if it is exposed to collision with water molecules quite often, then the

TABLE 3.1
Extinction Coefficients in Proteins

Residue	ε at 280 nm (liter/mmole cm)
Tryptophan (W)	5,500
Tyrosine (Y)	1490
Phenylalanine (F)	1
Cystine	125

energies of the electron orbitals of those residues will not be normal. The ability to absorb can be greatly distorted if there is no energy gap between the electron orbitals that would correspond to a photon with a wavelength of 280 nm. Thus, there are occasions even for very pure proteins when the calculated extinction coefficient is not the real one. This will skew the measured concentration.

Second, it is not true that albumin is a good standard for determining the concentration of all proteins by this method. Not all proteins have the same proportion of tryptophan, tyrosine, phenylalanine, and cysteine relative to their mass that albumin has. Some are much higher, and so a milligram of them would absorb more. This method would overestimate the concentration of such proteins. Likewise, a protein that was lacking absorbing residues might be very abundant, but would not absorb very well. It would be underestimated.

Third, there is a far more practical consideration: cost. This method requires the use of quartz cuvettes. Regular glass or plastic cuvettes are quite opaque to light at 280 nm, and cannot be used. Though they allow visible light to pass through readily, they are not much better than rocks at transmitting ultraviolet light. Unfortunately, quartz cuvettes can cost several hundred US dollars, which is cost prohibitive for some laboratories. Likewise the deuterium lamp is a very expensive piece of equipment and not easily available to every lab. If it were a perfect world, this would not be a consideration to scientists, but in the real world, a scientist trying to establish a laboratory may not be able to start out using this technique, no matter how ideal it would be otherwise.

2. THE BRADFORD ASSAY

The Bradford assay is one of the oldest, most reliable, and least expensive ways to determine protein concentration. It is rapid, and easy to do with a standard visible wavelength spectrophotometer.

The Bradford assay uses a particular chromophore to adhere to certain residues of a protein, forming a bright-blue complex. Bradford reagent when in complex with lysine, arginine, and histidine under acidic conditions will change color, and the color change can be detected spectrophotometrically. As with other spectrophotometric methods, a standard must be used, and that standard is generally albumin. The color change is proportional to the concentration of affected residues, not the milligrams of protein. If the proportion of such residues per milligrams is the same in all proteins as it is in albumin, then the color change can be normalized to the milligrams per milliliter of protein.

Unfortunately, the Bradford assay has the same weakness that the Abs280 method does: some proteins do not have the same ratio of residues per milligram as albumin; therefore, the actual concentration of protein in mg/mL may be quite different in your sample as it is in the standard.

3. BICINCHIONIC ACID ASSAY

Bicinchionic Acid (BCA) is a copper-based organic dye, which complexes to peptide bonds. As it does so, it also tends to unfold protein structure, such that all the peptide bonds in the protein backbone are exposed and complexed. When BCA complexes to a peptide, it changes color maximally at 582 nm. The color change is proportional to the concentration of peptides, and can be monitored spectroscopically.

In this case, the weakness regarding the standard is almost entirely eliminated. The method makes only the assumption that the proportion of peptides to residues is the same between the standard and the protein being monitored. Since, in a protein with "n" residues, the number of peptides is "n-1", the assumption is more than 99% true for any protein with more than 100 residues.

The worst difficulty with this method is that BCA reagent is both expensive and unstable. It undergoes the color change even in the absence of protein. The reagent must be prepared immediately before being exposed to the protein, and used within 30 minutes. Any leftover material is wasted. Beyond that, though, the only difficulty people tend to experience is accidentally confusing the BCA with the standard bovine serum albumin (BSA). Experience quickly deals with that difficulty, and sufficient funding deals with the cost.

FINDING THE CONCENTRATION

Determining the concentration of a protein is always done in two parts: making a standard curve then measuring the actual concentration via some dilutions of your protein sample.

To make a standard curve, you need to prepare appropriate dilutions of BSA. You do this by mixing a stock solution of BSA with different amounts of deionized water. If the stock concentration of BSA is 1.0 mg/mL, then the relative proportions of stock to water in order to make 100 μL are shown in the table below:

Desired Conc (mg/mL)	Volume of Stock BSA (μL)	Volume of Water (μL)
1.0	100	0
0.9	90	10
0.8	80	20
0.7	70	30
0.6	60	40
0.5	50	50
0.4	40	60
0.3	30	70
0.2	20	80
0.1	10	90
0.0	0	100

If you do not have 1.0 mg/mL BSA, then you will have to adjust your volumes accordingly in order to achieve the desired concentrations. Also, if you are going to be making dilutions of your protein for the following part, it is recommended you do so now (see second goal).

Dilution Factor: The value by which the measured concentration of a protein must be multiplied in order to determine the true concentration. *e.g. 100 μL of a protein is combined with 400 μL of buffer for a total volume of 500 μL. The protein is 1/5th as concentrated as it was before, and thus has a dilution factor of 5. If the measured concentration is 0.25 mg/mL, the actual concentration is 1.25 mg/mL.*

You must combine the standard concentrations with any dye reagent if you are using it (Bradford's or BCA, for example). You must then determine the absorbance, and plot absorbance versus the known concentration of your standard. As long as the absorbance does not go beyond 2.00 absorbance units, the response tends to be linear. You should mathematically determine it. An example of a standard curve made for the BCA test is shown in Figure 3.1.

The chart in Figure 3.1 indicates that there is a linear relationship between absorbance and concentration.

$$\text{Conc(mg/mL)} = (\text{Absorbance} - 0.0173)/1.4582$$

This is exactly the kind of relationship that is required in order to complete the second goal. You must generate a standard curve with BSA before you can proceed onwards.

In getting to the second part – finding the concentration of samples – you should make some dilutions of your protein. This is generally necessary because

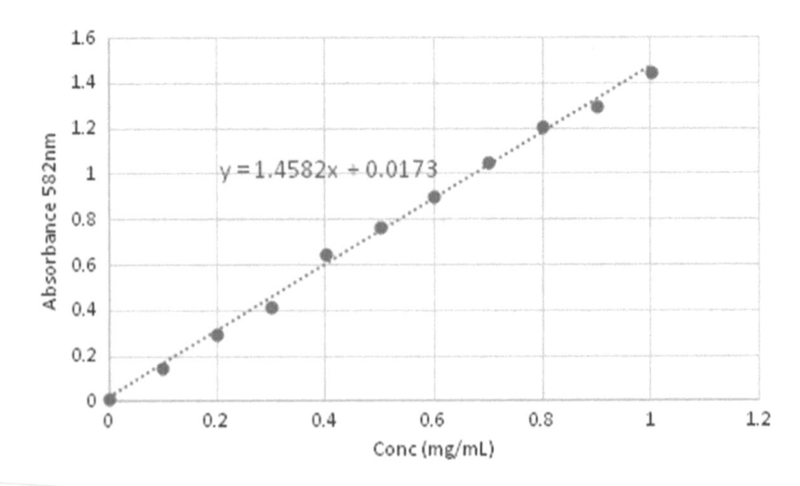

FIGURE 3.1 A standard curve of absorbance versus concentration for the BCA assay. The absorbance is directly proportional to concentration values, as seen here.

you do not know what the concentration is, and because you want to have an absorbance value less than 2.00, and preferably within the range of values on the standard curve. Since you do not know in advance which dilution will give such a result, you need to make a range of concentrations and find out which is the right one. The first time you ever work with a protein sample, you will want to make samples with dilution factors of 2, 5, 10, 20, 100, and 1000. For this exercise, that is what you will do. Later in your career, when you are working with other samples, you will have an idea already what the concentration is, and can make fewer dilutions.

If you are using a dye reagent, always use the same one as was used to make the standard curve, and make it at the same time you make the samples for your standard curve. Then, determine the absorbance of all the dilutions, and calculate the concentration.

e.g. Assume the standard curve from part 1 was made for this experiment. The sample with the dilution factor of 5 has an absorbance of 0.681. Its concentration would be:

$$\text{Conc(mg/mL)} = (0.681 - 0.0173)/1.4582 = 0.455\ \text{gm/mL}$$

However, since the dilution factor is 5, the actual concentration of the protein is 5 * 0.455 mg/mL, or 2.275 mg/mL. For all dilutions which give absorbances within the range of the standard curve, the calculated concentrations tend to be the same, as it should be. However, for samples whose diluted concentrations are much higher or much lower than the values in the standard curve, the calculated concentration of the original sample will deviate by large amounts. Such values are not to be trusted. A table below illustrates this principle:

Dilution Factor	Absorbance	Calculated Concentration
1	2.405	1.637
2	1.702	2.311
5	0.681	2.275
10	0.349	2.274
20	0.184	2.286
100	0.046	1.968
1000	0.011	–4.382

Notice how the values whose absorbance values are closest to the center of the standard curve tend to be grouped most closely. These are the most reliable values. Ones at the outer edge deviate further away from the true value, in some cases – such as the one with the dilution factor of 1000 – being obviously unreliable. When performing the concentration assay, you only should calculate concentrations from absorbance values that fall between your minimum and your maximum absorbance measurements for your standard curve.

Goal

In this lab period, you will develop a standard curve for protein concentration from BSA, and then determine the concentration of a sample of protein.

First Goal: Generate a standard curve to determine protein concentration.
Second Goal: Determine the concentration protein in extracts of beef liver, fish muscle, and carrot root, using both the standard curve and the known dilution factor.

Materials

Bradford reagent (see following recipe)
BSA stock solution 1.0 mg/mL
0.15 M Sodium Chloride
1.5 mL Eppendorf Tubes
An Eppendorf Tube rack to hold 45 tubes
Three samples of protein in 0.02 M phosphate buffer pH 7.4:

5 g beef heart extracted in 75 mL buffer
5 g fish muscle extracted in 75 mL buffer
5 g carrot root extracted in 75 mL buffer

0.02 M Phosphate buffer pH 7.4
A VIS spectrophotometer
A 1-mL VIS cuvette

Instructions

1. Label six tubes for your standards: "Blank", "0.05", "0.10", "0.15", "0.20", and "0.25".
2. Add volumes as indicated in the following table to make the desired concentrations listed. You will use the BSA stock and the NaCl solutions mentioned above to make these volumes:

Label	**Desired Conc (mg/mL)**	**Vol Stock (μL)**	**Vol NaCl (μL)**
Blank	0 mg/mL	0	100
0.05	0.05 mg/mL	5	95
0.10	0.10 mg/mL	10	90
0.15	0.15 mg/mL	15	85
0.20	0.20 mg/mL	20	80
0.25	0.25 mg/mL	25	75

3. Obtain thirteen Eppendorf tubes for each sample (total 39 tubes). For each set, label seven as "1:1", "1:5", "1:10", "1:50", 1:100", "1:500", and "1:1000". Label the other six as "1:5d", "1:10d", "1:50d", 1:100d", "1:500d", and

"1:1000d". Add the designator "B" for the beef liver tubes, "F" for the fish muscle, and "C" for the carrot root. These are your "sources" for enzymes.

4. Prepare dilutions of your unknown enzyme by adding volumes shown in the following table. To make these dilutions you will use the enzyme sample for some and some of the dilutions for others, in what is called a "serial dilution". The remaining volume will come from the phosphate buffer. Make these dilutions in the tubes that have the "d" designation for "dilutions".

Label	Dilution Factor	Source	Volume Source (μL)	Volume Buffer (μL)
1:5d	1:5	Enzyme sample	200	800
1:10d	1:10	Enzyme sample	100	900
1:50d	1:50	1:5d	100	900
1:100d	1:100	1:10d	100	900
1:500d	1:500	1:50d	100	900
1:1000d	1:1000	1:100d	100	900

It is much more accurate to prepare dilutions this way than to try to take all volumes from the original enzyme sample. In theory it should be possible to make a 1:1000 dilution by taking 0.1 μL of the enzyme sample and adding it to 99.9 μL of buffer. However, if your pipet is miscalibrated or even has an extra 0.1 μL residually clinging to the side, then your error will be enormous.

5. From each tube with the "d" designation, pipet 100 μL of the enzyme sample into the correspondingly labeled tube without the "d". For example, you would put 100 μL of "1:5d" into the "1:5" tube, and 100 μL of "1:10d" into the "1:10" tube, and so forth. You put 100 μL of the undiluted extracts into the tubes with the "1:1" labels.
6. Pipet 1 mL of Bradford Reagent into each tube that has a 100 μL aliquot. This includes every all of your BSA standards and all of the tubes that lack the "d" designation, for a total of 27 tubes.
7. Allow the tubes to stand at room temperature for 2 to 10 minutes.
8. Auto zero the absorbance at 595 nm against the 0gm/mL blank standard. Then, measure the absorbance at 595 nm for each BSA standard, beginning with the lowest concentration, and going upward. If you go from lowest to highest concentration and empty the cuvette every time, you will not have to clean the cuvette. Otherwise, clean the cuvette after every measurement. DO NOT DISCARD THE BLANK.
9. Generate your standard curve, using the absorbances of the samples. Find the line of best fit.
10. If your cuvette has been stained by Bradford reagent, clean it and dry it at this point or acquire a new cuvette. Auto zero the absorbance at 595 nm again with the blank from the standards.
11. Measure the absorbance of each of the dilutions of the beef liver extract, starting with the most dilute and going to the least dilute.

12. Clean your cuvette again, or replace it, and auto zero the spectrophotometer again with the blank standard.
13. Measure the absorbance of the dilutions of the fish muscle extract, as you did in step 11. Repeat step 12 again.
14. Measure the absorbance of the dilutions of the carrot-root extract, as before.
15. Calculate the concentration of each of your samples using the line of best fit, the absorbance, and the dilution factor. Only calculate the concentration using absorbances that lie in-between your minimum absorbance for a standard and your maximum accepted absorbance. Such concentrations will always be the most accurate.
    (Note: sometimes, a phenomenon known as the "inner filter effect" will occur with the highest concentration standard. In such cases, the absorbance of this standard is not proportional to the concentration, and can be seen significantly deviating to the right of the line of best fit in a plot of absorbance versus concentration. If this is observed, do not accept the absorbance from this standard as a true value. You should exclude it from your line of best fit when determining the equation that relates absorbance and concentration.)
16. Identify which source is the richest in protein.

Recipe for Bradford Reagent

To make 1 liter of Bradford reagent, add 50 mg of Coomassie Brilliant Blue G-250 to 50 mL of neat methanol and swirl to dissolve. Add 100 mL of 85% (w/v) phosphoric acid, H_3PO_4. Slowly pour the acid solution of dye into 850 mL of deionized water, letting any precipitating dye dissolve before continuing to add more of the acid solution. Filter the mixture using Whatman #1 filter paper, to remove any kind of precipitate that does form. Store the mixture in brown glass at 4°C until use.

PRE-LAB QUESTIONS

1. Suppose you have four standards whose concentrations are 0.05 mg/mL, 0.10 mg/mL, 0.15 mg/mL, and 0.20 mg/mL, and they absorb at 595 nm when 100 μL are mixed with 1 mL of Bradford reagent with absorbances of 0.249, 0.501, 0.744, and 0.963. You take a sample of protein, and you mix 150 μL of this protein with 850 μL of buffer, then take 100 μL of this dilution and mix it with 1 mL of Bradford reagent. The protein in this dilution has an absorbance of 0.810. What is its concentration?

2. In the preceding question, when the protein is used in undiluted form and 100 µL is mixed with 1 mL of Bradford reagent, the absorbance is measured as 1.422. What concentration is indicated by this absorbance, and why should you not believe that this is the true concentration?

POST-LAB QUESTIONS

1. If you have performed an activity assay lab for these samples in another experiment, do you see any relationship between the concentration of any enzyme and the total concentration of the protein? Should you? Explain your answer.

2. Using a website such as www.uniprot.org or some other source to provide this information, find the primary structure bovine serum albumin and human hemoglobin. Based off of the proportion of certain residues in each protein, determine whether the Bradford assay will overestimate, underestimate, or accurately determine the concentration of hemoglobin, when albumin is used as a standard.

3. Explain why the most accurate concentration is determined from dilutions that give a response within the range of absorbances shown by the standards as well.

4 ELISA

Enzyme-Linked ImmunoabSorption Assay, or "ELISA", allows an enzyme that can be monitored to indicate the presence of a non-enzymatic protein that cannot. Consider a common protein we have worked with in a previous chapter of this manual: albumin. What does it do? It does not have any enzymatic activity, and cannot be monitored visually. It does not have a spectroscopic absorbance profile that is significantly different than all other proteins. How can one quantify it, assay for it, or even locate it? There are many proteins like albumin, which have important biological roles, but which cannot be assayed for by the techniques we have discussed so far. Nonetheless, it is possible that you will have to locate and quantify these proteins.

ELISA uses the binding of antibodies to reveal the presence of target proteins, as shown in Figure 4.1.

First, an immunoglobin antibody binds to some *epitope* on the target protein. An epitope is just some feature on the surface which is recognized by the immunoglobin. After any immunoglobin that did not bind is washed away, a secondary antibody which recognizes the first antibody binds to it, and all antibody which did not bind is washed away. This second antibody is fused to a reporter protein. This reporter protein is visible, either by intrinsic color, or by some enzyme activity. Common reporter proteins include green fluorescent protein and horseradish peroxidase. Radioactive isotopes on the secondary antibody are also used sometimes, though this is less commonly done these days than it was in the past. The sequence of using antibodies in this order will only lead to the reporter protein giving a signal if the original target protein is present. Otherwise, the first antibody will not bind to anything, which then causes the second antibody not to bind to anything. Antibodies that do not bind are washed away, and this keeps the reporter protein from being present to give its signal.

Experimental considerations include the need to get the protein of interest to adhere to the walls of the reaction chamber. Microwells of polystyrene are usually used, because most proteins bind with some affinity to the polystyrene. However, it also creates the need to prevent the antibodies from randomly binding to the polystyrene walls, instead of to the target protein. Any such binding would lead to a false positive signal, and lead a scientist to conclude that the target was in higher abundance than it actually is. For this reason, it becomes necessary to "block" any exposed polystyrene surface area before the addition of any antibody. Blocking merely means saturating the surface with something else that the antibody will not bind to. Common and inexpensive proteins can be added in great abundance to saturate the polystyrene walls. The protein casein can be extracted from milk cheaply and easily, and is a popular choice, as is serum albumin. A 10% solution of the chemical "Tween 20", which is really polysorbate, also works to block the polystyrene, as does dried and powdered non-fat milk. Tween 20 is faster and more efficient, but also more expensive than casein or milk.

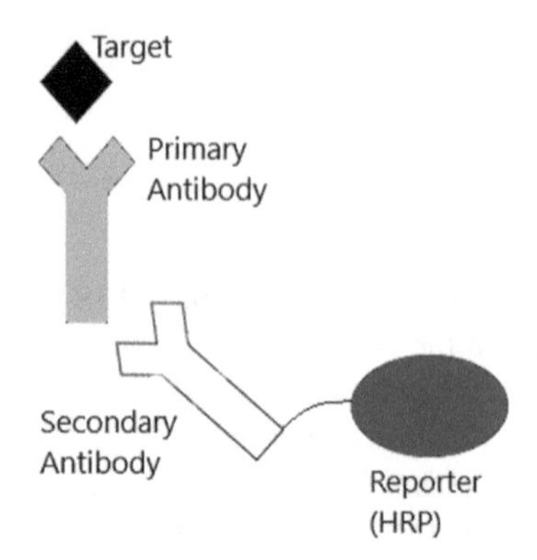

FIGURE 4.1 In ELISA, a target epitope is bound by a primary antibody, then a secondary antibody with a reporter protein binds the first antibody. The presence of the epitope is visualized by the activity of the reporter protein.

Other considerations include the method used to image, as has been mentioned. If the activity of horseradish peroxidase is used as the imaging device, its substrate – 2,2′,5,5′-tetramethyl benzidine (TMB) – produces a product which is blue and absorbs strongly at 652 nm. However, this product is also a pH indicator whose acid form is yellow, absorbing strongly at 450 nm. The horseradish peroxidase reaction, shown in Figure 4.2, can be quenched if the pH is lowered, but this will also change the product to its acid form and require visualization at a different wavelength.

A third consideration has to do with the time required. This laboratory has been written in such a way that results can be obtained within a four-hour time period. However, best results are obtained if longer time periods are used. For example, incubating the target material with the polystyrene or blocking the polystyrene with casein give the best results if done overnight, instead of 20 minutes. This laboratory attempts to compensate for the time gap by using a higher concentration of casein than would be used if done overnight, but the higher concentration also has its disadvantages. Some of the target protein can be displaced by the blocking reagent, especially if the target protein is small and has less surface area. If the target protein is displaced, it gives a false negative response. Using Tween 20 instead of casein somewhat ameliorates this problem, because it adheres more rapidly, but there is no real substitute for the longer time period if a researcher is trying to obtain very clear results.

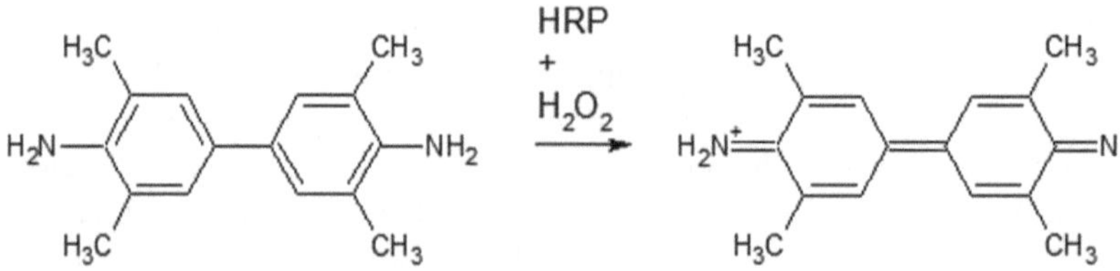

FIGURE 4.2 Horseradish peroxidase converts a colorless precursor into a colored product, in conjunction with the reduction of hydrogen peroxide.

Because ELISA is merely another assay technique, it is in some ways thematically connected to the previous chapter on assays. It is a more widely used tool in molecular biology, however, and will later on be connected to a technique that separates proteins from each other. Its utility as an assay is strongest in combination with other methods. To use it in other methods, like the enzymatic assays, it must be first mastered as its own separate skill.

GOAL

The purpose of this experiment is to determine the presence and relative abundance of collagen in three different sources.

MATERIALS

1 gram of bovine liver homogenized in 100 mL of 0.02 M phosphate buffer, pH 7.4
1 gram of bovine bone homogenized in 100 mL of 0.02 M phosphate buffer, pH 7.4
1 gram of carrot root, homogenized in 100 mL of 0.02 M phosphate buffer, pH 7.4
5 grams of casein, dissolved in 100 mL of 0.02 M phosphate buffer, pH 7.4
Primary antibody, mouse IgG anti-bovine collagen, 5 μg/mL in 100 mM carbonate/bicarbonate buffer, pH 9.6

(N.B. any mouse-derived antibody against any protein of interest will work in this experiment, except an antibody against casein.)

Secondary antibody, rabbit anti-mouse antibody, conjugated to horseradish peroxidase, 5 μg/mL in 100 mM carbonate/bicarbonate buffer, pH 9.6
2,2′,5,5′-tetramethyl benzidine (TMB), 5 μM in deionized water
2 M H_2SO_4 (optional)
Polystyrene microplate wells

INSTRUCTIONS

1. Pipet 50 μL of each homogenate into three different polystyrene wells. Label the wells "L" for bovine liver, "B" for bovine bone, and "C" for carrot root. Pipet 50 μL of 0.02 M phosphate buffer, pH 7.0 into a fourth well as a blank, and label it " – ". Allow the wells to incubate at 37°C for 30 minutes. The proteins in the homogenate should bind to the walls of the well during this period.
2. Pour out the microplate wells into a waste beaker or onto a paper towel, being careful not to splash the contents of any well into another. Tap the wells once or twice to remove excess droplets.
3. Wash the wells by filling them with the solution containing casein. Be extremely careful that you do not overfill the wells so that they spill into each other. Incubate at 37°C for 15 minutes to allow casein to bind to every exposed position on the polystyrene surface.

4. Once again, pour out the microplate wells into a waste beaker or onto a paper towel. Then repeat the wash step again, and pour out the contents again.
5. Pipet 50 μL of the mouse anti-bovine collagen antibody, which will be the primary antibody. Allow it to incubate at 37°C for 30 minutes. This should give time for the antibody to complex with any collagen that is present and bound to the polystyrene surface.
6. Pour out the microplate wells again, and wash with casein as before. Be sure to wash it two times, so that any unbound antibody is removed.
7. Pipet 50 μL of the rabbit anti-mouse IgG antibody into each well. Allow it to incubate at 37°C for 30 minutes. This gives time for the secondary antibody to complex with the primary antibody.
8. Pour out the microplate wells as described above and wash with casein two times so that all the unbound secondary antibody is removed.
9. Pipet 50 μL of TMB solution into each well. Cover the wells with a paper towel so that most of the ambient light is blocked out. Incubate at 37°C. Allow the horseradish peroxidase 15 minutes to convert the substrate into its colored product.
10. Add 50 μL of 2 M H_2SO_4 in order to stop the HRP reaction. (This is optional, but if you do not do it, you must work quickly in the next step, especially if you are imaging it.)
11. Observe the wells, and rank them from least abundant in collagen to most abundant.

    A vertical plate reader can be used to quantify numerically the absorbance of the product at 450 nm. Any camera that allows numerical quantification of pixellation at 450 nm can also be used to achieve this result, though not as accurately. It is mentioned here, however, because most modern phones include such cameras and software that quantifies pixellation. Not all biochemistry labs have plate readers. If you did not add 2 M H_2SO_4 then you should image at 652 nm.

PRE-LAB QUESTIONS

1. What is the function of collagen?

2. Why do you use a solution of casein to wash, rather than wash with just the buffer?

3. If an epitope is not present in the well, what will happen to the primary antibody and to the secondary antibody?

4. In this lab, the reporting protein is horseradish peroxidase, conjugated to the secondary antibody. Locate and draw the structure of the substrate for horseradish peroxidase, TMB, and the structure of the product into which it is converted.

POST-LAB QUESTIONS

1. What would be the likely result if the washing steps had been omitted?

2. Which sample was richest in collagen? Explain the biological reason why this is the case.

3. Why was a negative control performed in this experiment? How are your results skewed if you do not perform a negative control?

4. Consider the primary antibody that you used. If you had used this antibody on carp jaw bones, you might have observed no color development when you added the secondary antibody and the substrate TMB. And yet, carp bones do have collagen, as do all bony and cartilaginous fish. Why might you still get a negative response?

5 Salting Out Proteins and Other Biomolecules

Organic chemists, especially relative novices at the craft, tend to think of molecules as "soluble" or "insoluble" in their totality. A molecule either extracts to hexane or else remains in an aqueous layer, with only some partitioning in between. The factors that make it soluble in one layer or another depend upon the overall polarity of the molecule. If it has a high proportion of hydroxyl groups to the number of carbons, it will probably partition mostly to the aqueous layer. Otherwise, it will prefer a non-polar solvent. However, more advanced organic chemists reach the understanding that polarity is not as much a "total molecule" phenomenon as a "localized region" phenomenon. Some parts of the molecule prefer to partition to an aqueous layer, other parts prefer a non-polar solvent, and the same molecule might well exist attracting both kinds of solvent molecules simultaneously.

The size of biomolecules means that all biochemists have to come to this understanding. A 500 amino acid residue polypeptide is going to have patches that are distinctly polar – even ionized – and patches that are quite non-polar. The polar parts will interact with water, and the non-polar parts will not. What then determines whether the biomolecule will be soluble in water?

Partially, the solubility is maximized by the folding of the protein. The protein is a polypeptide of amino acid residues, each of which has a certain degree of polarity or hydrophobicity. Most globular proteins fold such that their hydrophobic residues are packed to the inside of the protein and their more polar residues mostly face outward. This arrangement excludes water from the hydrophobic residues and maximizes the interaction with water for the hydrophilic ones. In total, this results in the protein being soluble: its residues interact with water well enough that the protein does not try to interact with itself. If it did, then the protein would precipitate out of solution.

That is the exact basis upon which the principle of "salting out" works. If conditions change such that the protein's hydrophilic residues do not interact with water, then it must fall out of solution. The only question is how much the conditions have to change in order to make any given protein precipitate. The ones that are interacting the least strongly would precipitate with mild stress, and the ones that interact the most strongly will require much greater stress in order to precipitate.

Let us take a very simple example in order to make clear how salting out works. Consider a protein whose outer surface has a number of hydrophobic patches and regularly spaced lysine and glutamate residues. This protein bears a marked resemblance to a soccer ball. Because the lysine and glutamate residues are charged, they strongly attract a shell of water molecules around them due to their electrostatic attraction. It is the large number of water molecules attracted to the surface of this protein that keeps it "dissolved" because they hydrophobic patches would rather be excluded from the water molecules. If anything were to strip the water molecules away, or remove the electrostatic attraction, the protein would not stay in solution.

A number of things might happen to dissociate the water molecules. Suppose the pH of the solution were to drop down to 2.0. Almost every glutamate residue would become protonated, and in the process lose its negative charge. When that occurs, the waters would no longer experience electrostatic attraction to the glutamate, and would dissociate. The proportion of this "soccer ball" protein's surface that had a shell of water molecules coordinated around it would diminish. Now, the hydrophobic patches on the surface of one protein molecule might exclude themselves from water by adhering to the surface of a nearby protein molecule. The hydrophobic patches would be buried, and the protein less likely to remain dissolved. Either the protein would refold itself completely in order to get more polar surface area exposed to the water, or else it will just precipitate out of solution, or perhaps both will occur. Whatever effect is observed is due to the coordinated shell of water molecules being lost from the protein.

Salting out would occur if salt was added to the solution in which this protein resides. The salt is composed of cations and anions. As it dissociates, its ions are going to attract water molecules to themselves. Every water molecule that is coordinating to an ion is NOT coordinating itself in a solvent sphere around the glutamate or the lysine. If there are many free waters in solution, this will have no effect on the protein. Waters that happen to dissociate from the protein would quickly be replaced by free waters in the solvent in a normal equilibrium process. However, as more salt is added, more water gets coordinated to the ions of the salt. At a certain point, water that dissociates from the protein will not be replaced by any free water in the solvent, because all the water molecules that were in the solvent are now coordinated to the ions of the salt. Possibly, the water will even get replaced by the cations and anions of the salt itself! At this point, the protein is no longer dissolved. It precipitates out of solution because there simply was not enough water to keep it in solution any longer. The addition of the salt drove the protein towards its saturation point, without changing the protein's concentration, by decreasing the amount of available solvent.

The biochemist who is going to salt his protein out of solution must make a choice regarding the salt to be used. Any salt will cause the salting out effect. Some salts, however, will also disrupt the structure of the protein. Lithium chloride, for example, produces lithium cations and chloride anions. Each of these ions has a very high charge density. That causes them to attract oppositely charged ions more strongly than another salt with the same charge but more diffused charge density. In our soccer ball protein, a glutamate residue might be bent to attract one of the lysine residues. As it bends, it may well put a bend into the entire backbone of the protein, and that bend might be vital for the functionality of the protein. Suppose the glutamate attracts a lithium with its high charge density. The glutamate will be pulled more strongly to the lithium than it was to the lysine. Indeed, the lithium might even cause a second glutamate to twist and be attracted to it. The change in the strain of the backbone could easily disrupt the whole structure of the protein, certainly precipitating it but also "pickling" it. The activity of the protein could be permanently lost, which is not what the biochemist intended. The same phenomenon would occur with use of the chloride or any other densely charged anion.

On the other hand, a salt whose charge was somehow diffused would be ideal for salting out. Ammonium sulfate is one such salt. The ammonium cation, NH_4^+, diffuses the positive charge such that only about a quarter of a positive charge is on any single hydrogen. Similarly, the sulfate, SO_4^{2-}, diffuses the negative charge such that each oxygen has approximately one half of a charge. Should the glutamates attract an ammonium or the lysines attract a sulfate, there will be minimal disruption of the backbone structure. The ion-to-ion attractive forces are not strong enough to disrupt the fold of the protein, except in the most unstable of polypeptides. Instead, the ammonium and the sulfate gently strip away water molecules until the protein precipitates. Generally, if more water is added, the protein goes right back into solution, recovering almost all of its previous activity. This makes ammonium sulfate an ideal salt for precipitations. Additionally, it is not strongly oxidizing or reducing, and is quite inexpensive to acquire in most cases. For this reason, the choice of other salts for the salting out process is rare, unless the protein is so stable that it does not readily unfold with any salt (*e.g. ribonuclease*).

Salting out is used in a number of biochemical techniques. It is the basis of many types of crystal growth experiments for any type of biomolecule. It is used to measure protein stability. A variation of it is used in protein unfolding experiments. Its most widespread use is in protein purifications, where it can eliminate two thirds of all protein impurities in a couple of hours' work.

When ammonium sulfate is used to purify proteins, it is added up to certain levels of its own saturation. At 0°C, a solution which has 0.697 g/mL of ammonium sulfate is 100% saturated (see Figure 5.1). If ammonium sulfate is added to a solution, the proteins that have the lowest ratio of water molecules to surface area will precipitate first. Those with the highest ratio will precipitate only when the ammonium sulfate concentration approaches 100%. Broadly depicted, one third of all cellular proteins will precipitate when ammonium sulfate is at 40% of its maximum saturation, and two thirds of all cellular proteins precipitate at 65% ammonium sulfate saturation. Thus, one can precipitate a third of the proteins, pellet them by centrifugation and separate them from the supernatant, then precipitate another third of the proteins as well, saving the pellet again. Each pellet would contain one third of all the soluble proteins, and the remaining supernatant would contain the last third. Your protein would thus be isolated from two thirds of all the other proteins from which you are trying to separate it.

Initial Concentration of ammonium sulfate % saturation	Final concentration of ammonium sulfate -% saturation at 0°																
	20	25	30	35	40	45	50	55	60	65	70	75	80	85	90	95	100
	grams of solid ammonium sulfate to add per milliliter of solution																
0	.106	.134	.164	.194	.226	.258	.291	.326	.361	.398	.436	.476	.516	.559	.603	.650	.697
5	.079	.108	.137	.166	.197	.229	.262	.296	.331	.368	.405	.444	.484	.526	.570	.615	.662
10	.053	.081	.109	.139	.169	.200	.233	.266	.301	.337	.374	.412	.452	.493	.536	.581	.627
15	.026	.054	.082	.111	.141	.172	.204	.237	.271	.306	.343	.381	.420	.460	.503	.547	.592
20	0	.027	.055	.083	.113	.143	.175	.207	.241	.276	.312	.349	.387	.427	.469	.512	.557
25		0	.027	.056	.084	.115	.146	.179	.211	.245	.280	.317	.355	.395	.436	.478	.522
30			0	.028	.056	.086	.117	.148	.181	.214	.249	.285	.323	.362	.402	.445	.488
35				0	.028	.057	.087	.118	.151	.184	.218	.254	.291	.329	.369	.410	.453
40					0	.029	.058	.089	.120	.153	.187	.222	.258	.296	.335	.376	.418
45						0	.029	.059	.090	.123	.156	.190	.226	.263	.302	.342	.383
50							0	.030	.060	.092	.125	.159	.194	.230	.268	.308	.348
55								0	.030	.061	.093	.127	.161	.197	.235	.273	.313
60									0	.031	.062	.095	.129	.164	.201	.239	.279
65										0	.031	.063	.097	.132	.168	.205	.244
70											0	.032	.065	.099	.134	.171	.209
75												0	.032	.066	.101	.137	.174
80													0	.033	.067	.103	.139
85														0	.034	.068	.105
90															0	.034	.070
95																0	.035
100																	0

FIGURE 5.1 The masses of solid ammonium sulfate to be added are shown to make a final % saturation, given the starting concentration. For example, if you have a 40% saturated solution and wish to make a 65% saturated solution, you would have to add a further 0.153 grams of ammonium sulfate for every milliliter of solution.

GOAL

Your goal is to separate a cellular extract into three portions of soluble protein and determine into which portion catalase and lactate dehydrogenase segregate.

METHOD

You will be given an extract of beef liver or fish muscle, known as a "crude". It will be in a light extraction buffer, probably 0.02 M phosphate pH 7.0. Keep it on ice whenever you are not using it. Determine the volume of the crude, and place it in a beaker. Calculate the mass of ammonium sulfate you must add to bring the solution to 40% saturation.

Slowly add the ammonium sulfate, stirring constantly. (**IMPORTANT:** you must add it only a few grains at a time and make sure that each addition is completely dissolved before adding any more. If you raise the *local* concentration of ammonium sulfate to higher than 40% saturation, you will get precipitation of proteins you do not want.) It should take about five minutes to add all the ammonium sulfate.

The mixture will have a precipitate suspended in it. Place this whole mixture in one or more clean centrifuge tubes. Balance the mass in the centrifuge tubes to within 0.1 g or less, either by exchanging quantities between two different tubes or by adding an insoluble and non-reactive substance such as boiling chips, glass marbles, or sand. Centrifuge at 12,000×g for 10 to 15 minutes. Consult the specifications of your rotor and centrifuge to determine how many RPM correspond to this centrifugal force. Pour the supernatant into a clean beaker. Resuspend the pellet in about 20 mL of buffer and pour it into a different beaker. If you added any insoluble component in order to balance the mass, you should discard it at this time, but be certain you save both the pellet and the supernatant. Be certain to label all containers.

Calculate the mass of ammonium sulfate you must add to bring the 40% saturated solution up to 65% saturation. As before, add this mass slowly with constant stirring. Place the mixture into clean centrifuge tubes, balance the mass, and centrifuge again at 12,000×g for 10 to 15 minutes. Pour the supernatant into a clean beaker. Resuspend the pellet in 20 mL of buffer of buffer, as before, and pour it into a different beaker. Once again, label your beakers as "40% pellet", "65% pellet", and "above 65% supernatant". These are the three fractions into which your proteins have segregated.

Assay all the fractions for your proteins. Determine the relative activity. **IMPORTANT:** Do not discard any fraction until after you have assayed everything and are certain where your protein is or is not.

In a protein purification, you would at this point discard the fractions that do not have your protein, and dialyze the fraction that does have your protein against a buffer. This is done in order to remove the salt. In this exercise, however, you will discard all fractions down the sink and clean up your station.

DATA ANALYSIS

In theory the total activity should not diminish. In practice, it often does. To analyze and explain the results, consider the answers to the following questions

a. Determine the initial total activity of each enzyme.
b. After addition of ammonium sulfate to the 40% level, your enzyme will be partitioned between the pellet and the supernatant. Determine the total activity of each enzyme in the pellet and what is the total activity of each enzyme in the supernatant. Determine whether the combined total activities between the pellet and supernatant equal the initial total activity.
c. If your total activity decreased between the initial and 40% level, explain why. Similarly, if it increased, explain why you saw this apparently impossible phenomenon.
d. Determine the total activity at the 65% level of your enzyme in the portion to which it partitioned (pellet or supernatant). Calculate the percentage of the initial total activity which you recovered.
e. Determine into which fractions the two proteins partitioned.

PRE-LAB QUESTIONS

1. What mass of ammonium sulfate do you have to add if you want to achieve 35% saturation in a 22 mL sample of protein, assuming you have added no salt yet?

2. What mass of ammonium sulfate do you have to add if you want to achieve 60% saturation in 22 mL of a sample of protein which already has 25% saturation?

3. If you add ammonium sulfate to the 40% level and initially have 268 total units of enzyme activity, but only recover 110 units of enzyme activity in the supernatant after centrifugation, there are two possible explanations for what may have happened to the missing 158 units. State them both.

POST-LAB QUESTIONS

1. When salting out from beef liver, neither the catalase nor the L-LDH is red. Catalase is very faintly brown, and L-LDH is completely clear in visible light. Which protein(s) is/are likely to be the red ones?

2. Which protein is more likely to partition into a 40% pellet: one enriched in hydrophobic residues at the surface, or one with many hydrophilic residues? Explain your answer.

3. This problem will require you to acquire some sort of three-dimensional viewer program. Several are available. One such is Cn3D, which can be acquired from the United States National Institutes of Health National Center for Biomolecular Information (www.ncbi.nlm.nih.gov/). You will also need to access the structural files of the enzymes that have been studied so far in this text. The files can be accessed at the same site above, and are given four character codes to uniquely designate them. Porcine L-lactate dehydrogenase (5LDH), carp L-lactate dehydrogenase (1V6A) are likely the closest available structures to what you have had available, because no structure of bovine L-lactate dehydrogenase has ever been solved, and very few piscine ones have been solved. Similarly, acquire the structure of bovine catalase (3RE8). No structure for catalase from any mammal has ever been solved, and the nearest match is gamma glutaryl hydrolase from the Zebrafish. It, however, has only 23% identity with catalase from the zebrafish, whereas bovine catalase has 83% identity, thus making it a better model for analyzing and predicting structural details.
   a. Having acquired the ability to view the structures, compare the structures of porcine L-LDH and carp L-LDH. Color the residues according to hydrophobicity. Rotate the structures in three-dimensional space to get an impression of the surfaces. Which one appears to be more hydrophobic, if either? Make a prediction as to which ammonium sulfate fraction they will segregate: the same fraction, the porcine L-LDH at lower ammonium sulfate concentration, or the carp L-LDH at lower ammonium sulfate concentration? Explain your answer.

   b. Now compare the structures of porcine L-LDH and bovine catalase, again coloring the residues according to hydrophobicity and looking at the structures from all sides. Make the same kind of prediction as to where they will segregate: the same fraction, the porcine L-LDH at lower ammonium sulfate concentration, or the bovine catalase at lower ammonium sulfate concentration? Explain your answer.

   c. How do your predictions in part (b) compare to your experimental results? Explain any discrepancy.

6 A Discussion of Isoelectric Point and Effective Charge

This section is not a lab, but is an important theoretical section, which you need to understand in order to design experiments based on ion exchange chromatography, native gel electrophoresis, crystal growth, salting out, and other experiments.

EFFECTIVE CHARGE

Calculating the effective charge on a protein is not difficult. Mathematically it is quite similar to calculating the charge on a dipeptide, with one extra step added in.

Consider first calculating the charge on a dipeptide: "What is the effective charge on the Gly-His dipeptide at pH 5.5?"

First of all, there are three ionizable groups in this structure: the N-terminus with pKa 9.60, the His side chain with pKa 6.0, and the C-terminus with pKa 1.82. All these groups can be described by the Henderson–Hasselbach equation:

$$pH = pKa + \log([base]/[acid])$$

Of course, the H-H equation is only an approximation, and only works within a couple of pH units of any given pKa. If we were at pH 5.5, then we are so much higher than the pKa of the C-terminus that effectively ALL of the carboxylate there is in the base form. The percentage is close to 99.99%, and we can generally ignore the effect of the extra 0.01%. Likewise, we are so far below the pKa of the N-terminus that effectively all of it is in the protonated acid form. The only pH we will have to consider is the one closest to pH 5.5.

So let us assume we are at pH 5.5, and we are going to have to consider the pKa of the histidine. Now the histidine-residue-protonated form of the structure has a net charge of +1, and the histidine-residue-unprotonated form of the structure has a net charge of 0. Let us call these two forms "A" and "B" and let our equilibrium be:

$$A \rightleftarrows H + +B.$$

Which is a normal acid equilibrium. The H-H equation is now

$$pH = pKa + \log(B/A),$$

and we can rearrange it so that:

$$\log(B/A) = pH - pKa.$$

We also know that pH = 5.5 and pKa = 6.0, so …

$$pH - pKa = -0.5.$$

This allows us to solve for the term we most need: the ratio of base to acid.

$$\log(B/A) = -0.5.$$

In order to figure out the net effective charge, we need to know what percentage is in the +1 charged state (A) and what percentage is in the uncharged (B) state. We do know that the total amount is going to be the sum of B and A, so

$$A + B = 1.$$

This in turn lets us figure out the amount of A in terms of B:

$$A = 1 - B.$$

We are going to use this value back in our derivative of the H-H equation. We left that at:

$$\log(B/A) = -0.5.$$

To extract B/A, take everything to the power of 10

$$B/A = 10^{-0.5}.$$

But we can plug in our definition of A …

$$B/(1 - B) = 10^{-0.5}.$$

And we can multiply both sides by the denominator.

$$B = 10^{-0.5} * (1 - B),$$

which becomes

$$B = 10^{-0.5} - (B * 10^{-0.5}),$$

after we distribute through the parentheses. Then we group all the "B" containing terms on the same side of the equation:

$$B + (B * 10^{-0.5}) = 10^{-0.5}$$

and we factor B out of the terms on the left:

$$B * (1 + 10^{-0.5}) = 10^{-0.5}.$$

Now we can isolate B by dividing...

$$B = 10^{-0.5}/(1 + 10^{-0.5}),$$

and this can be solved by simply calculation:

$$B = 0.24,$$

or "24% of the dipeptide is in the uncharged B form".

Therefore, the remaining 0.76, or 76% must be in the +1 charged A form (which we expect, since the pH is less than the pKa).

At any rate, the effective charge is the weighted average of the charges, and we just found the weights.

$$\text{Eff. Charge} = (0.76 * +1) + (0.24 * 0) = +0.76$$

The peptide is somewhat charged, with a net +0.76 over the POPULATION. Any individual protein will be +1 or 0, but the population as a whole is showing a net average of +0.76.

ISOELECTRIC POINTS

Now, moving on to the only slightly more difficult case of the isoelectric point. In this case, there are two equilibria, not one. This is analogous to having the carboxyl and amino groups on a single amino acid, perhaps glycine. Both residues are on the same peptide, and are part of a linked equilibrium. Thus, we could write the linked equilibrium as:

$$H_2A^+ \rightleftarrows H^+ + HA \rightleftarrows H^+ + A^-$$

There are two equilibria:

$$K_1 = [H^+][HA]/[H_2A^+],$$

and

$$K_2 = [H^+][A^-]/[HA].$$

The isoelectric point would occur when $[H_2A^+]=[A^-]$. Since in a linked equilibrium, the combined equilibrium has an equilibrium constant that is the product of the equilibria that are added together, the equilibrium

$$H_2A^+ \rightleftarrows 2H^+ + A^-$$

would have

$$K_3 = [H^+]^2[A^-]/[H_2A^+] = K_1 * K_2.$$

This expression can be seen logarithmically, to bring it into comparison with pKa values.

$$-\log(K_3) = -\log([H^+]^2[A^-]/[H_2A^+]) = -\log(K_1 * K_2).$$

Just as with the H-H equation, this can be rearranged:

$$(2 * -\log[H^+]) - \log([A^-]/[H_2A^+]) = -\log K_1 + -\log K_2.$$

And certain substitutions can be included now:

$$2 * pH - \log([A^-]/[H_2A^+]) = pK_1 + pK_2.$$

This equation can be rearranged to isolate the pH term:

$$pH = \{(pK_1 + pK_2)/2\} + \log([A^-]/[H_2A^+])/2$$

Notice that we distributed the "2" in the denominator when we divided. This was to allow a further simplification, and the creation of the "pI" term. At the isoelectric point, $[A^-]=[H_2A^+]$, so $\log([A^-]/[H_2A^+])=0$. The logarithmic equation simplifies to:

$$pH = pI = \{(pK_1 + pK_2)/2\}$$

at the isolectric point. It is important to recognize that these pK_a values are the ones that flank the −1 charged form and the +1 charged forms. If you have a peptide of the sequence

Met-Arg-Glu-Ser-His

there are five ionizable groups with pKa values 1.82, 4.25, 6.00, 9.21, and 12.48. In its most positive form, there are three separate +1 charges on the peptide for a total positive charge of +3. Each pK_a value oversees an equilibrium where the charge drops by 1, so after pH 4.25, the charge will be +1 and after 9.21, the charge will be −1. pK_1 and pK_2 values referred to earlier. In general, for a peptide which has a charge of "+n" in its most positive form, the pK_a values used to calculate the pI will be the "nth" pK_a value and the "nth +1" value, 6.00 and 9.21 in this case.

Back to calculating the net charge, using the pI, at any other pH than the pI,

$$pH = pI + \log([A^-]/[H_2A^+])/2$$

This was the equation we needed to get to in order to determine what the proportion of $[A^-]/[H_2A^+]$ equals at any pH, and now it can be done almost the same way as in a simple single equilibrium step as shown earlier.

$$[A^-]/[H_2A^+] = 10^{\{2*(pH\text{-}pI)\}}$$

Suppose the pI = 6.0 and the pH = 6.4. The proportion of anion to cation is

$$\begin{aligned} [A^-]/[H_2A^+] &= 10^{\{2*(6.4-6.0)\}} \\ &= 10^{(0.8)} \\ &= 6.31 \\ [A^-] &= 6.31 * [H_2A^+] \end{aligned}$$

But for proportion, we also know that $[A^-] + [H_2A^+] = 1$, since they must sum up to the totality of the ion forms, *i.e.* "1" (NB: as stated before, this is only mostly true, since the miniscule effect of other equilibria does exist, but such effects are negligibly small).

We can substitute

$$[A^-] + [H_2A^+] = 6.31 * [H_2A^+] + [H_2A^+] = 1,$$

and isolating $[H_2A^+]$

$$[H_2A^+] = 1/(6.31 + 1) = 0.137,$$

or 13.7% of it is in the cationic H_2A^+ form. The remaining 86.3% is in the anionic form. This allows us to determine the charge as a weighted average:

$$(0.137 * {}^{+}1) + (0.863 * {}^{-}1) = -0.726.$$

And −0.726 is the effective charge on the protein. No single protein will actually bear this charge, but the population as a whole will average out to this, and single proteins will be fluctuating between charge states constantly, and so quickly that each protein will behave as if this was the charge it bore.

So it is not very different than calculating the charge in a more simple peptide. It is slightly more difficult to calculate with certainty the charge on a larger protein. The proximity in 3D space of one residue to another one will reciprocally affect the pKa values of the groups. However, for a first approximation of the pI, this technique works well, and is used to design experiments.

Amino Acid Name	α Carboxyl Group	α Amino Group	Ionizable Side Chain
Alanine	2.34	9.69	N/A
Glycine	2.34	9.60	N/A
Isoleucine	2.36	9.60	N/A
Leucine	2.36	9.60	N/A
Methionine	2.28	9.21	N/A
Phenylalanine	1.83	9.13	N/A
Proline	1.99	10.60	N/A
Valine	2.32	9.62	N/A
Asparagine	2.02	8.80	N/A
Glutamine	2.17	9.13	N/A
Serine	2.21	9.15	N/A
Threonine	2.09	9.10	N/A
Tryptophan	2.83	9.39	N/A
Aspartic Acid	1.88	9.60	3.65
Glutamic Acid	2.19	9.67	4.25
Histidine	1.82	9.17	6.00
Cysteine	1.96	10.28	8.18
Tyrosine	2.20	9.11	10.07
Lysine	2.18	8.95	10.53
Arginine	2.02	9.04	12.48

QUESTIONS

1. What is the charge on the peptide Met-Val-Leu at pH 7.0?

2. What is the isoelectric point of the tripeptide Ala-Gly-Cys?

3. The tetrapeptide Met-Trp-Arg-Glu has pI 6.73. What is the effective charge on this peptide in a TRIS buffer at pH 8.0? Phosphate buffer at pH 7.0? Imidazole buffer at pH 6.0?

4. A species of cone snail (*T. variegata*) secretes a peptide with the sequence: TRICCGCYWNGSKDVCSQSCC. Identify the isoelectric point, and determine the pH at which the peptide will have the effective charge of −0.50.

5. (Prior to performing ion exchange chromatography.) Your instructor will assign you a protein to purify. Determine the isoelectric point of the protein. Then determine the pH at which you should bind the protein to the column (net charge of −1) and release the protein from the column (net charge −1).

7 Column Chromatography

Chromatography is the separation of components over time. Originally, the technique was used to separate colored components, which is the origin of the name – "chromatos" means "color" in Greek. Modern chromatographic science can use a wide variety of retention methods, allowing separation of virtually anything under the right conditions, and isolation of it afterwards.

Chromatography works based upon the principle of "relative affinity", or "how tightly your chemical wants to stick to one substance compared to another one". Biochemically, the chemicals being separated are often proteins. Column chromatography is one of the best purification tools for proteins that can be used. Multiple types of column chromatography exist, and they have only a few similarities. One similarity, of course, is that they are all performed using materials flowing through a glass or plastic column.

There are always two components to a chromatography device. One is called the "stationary phase". It is a substance that does not move anywhere, and is commonly just paper, powdered cellulose, powdered plastic of some kind, or silica gel. Even normal cement can sometimes be used as a stationary phase. The solid-phase material is packed inside a column, from which it does not leave. The type of stationary phase material being used determines the kind of chromatography that separates a group of proteins. The other component to chromatography device is called the "mobile phase" and it is a substance that DOES move, and flows past the stationary phase in one direction. Common mobile phases are water, alcohol, non-polar solvents like hexane, gasoline, or acetonitrile. Sometimes gaseous substances like nitrogen, argon, or helium are used. In biochemistry, the mobile phase is most commonly some buffer or mix of buffers, because proteins and other biomolecules exist in buffered aqueous conditions. The researcher collects fractions of the mobile phase – usually in test-tubes of equal or similar volume – and assays these fractions for the protein of interest. Most other proteins will have separated to different fractions than the ones that contain your protein.

To separate any two proteins, A and B, they have to have different RELATIVE affinities for the stationary and mobile phase. Suppose that protein A sticks to the stationary phase with a value that we will call "1" and the mobile phase with a value which we will call "2". The relative affinity is ½. This means that it is half as likely to stick to the stationary phase as the mobile phase, and will move along reasonably quickly with the mobile phase. Suppose that protein B sticks to the stationary phase with a value of "3" (as in "three times more strongly than protein A"). That knowledge alone will not tell us how quickly it moves with the stationary phase, because we need to know the RELATIVE affinity, not the ACTUAL affinity. If protein B sticks to the mobile phase with a value of "2", then it has a relative affinity of 3/2, and is three times more likely to stick to the stationary phase than the mobile one. It will therefore move very slowly with the mobile phase. Not very much time will pass before protein A will have moved much further along than protein B, and they will be separated.

Essentially, each protein dissociates the stationary phase with a weak equilibrium constant, K.

$$A \bullet \text{stat.phase} \rightleftarrows A + \text{stat.phase}, \quad K_1 = [A][\text{stat.phase}]/[A \bullet \text{stat.phase}]$$

$$B \bullet \text{stat.phase} \rightleftarrows B + \text{stat.phase}, \quad K_2 = [B][\text{stat.phase}]/[B \bullet \text{stat.phase}]$$

If K_1 is larger than K_2 then it must be shifted toward the products more when A is dissociating from the stationary phase than when B is dissociating. Since A moves with the mobile phase whenever it is not bound to the stationary phase, it must move more often. Consequently, it moves faster. In practice, this technique works well when both K_1 and K_2 are fairly weak, but they must be noticeably different from each other.

As said before, the type of material comprising the solid matrix determines the kind of chromatography that occurs. The solid phase offers a chemical reason why a protein would want to associate with it. Whatever that chemical reason is, creates the basis for the value of K. There are three broad classes of chromatographic materials, though each broad class has many subclasses.

TYPES OF CHROMATOGRAPHY

1. **Ion exchange chromatography** (IEX) separates proteins based upon their isoelectric points. In this case, the solid phase has a functional group attached to it which has a distinct charge. Examples are quaternary amine groups to create permanent positive charges or else carboxymethyl groups or phosphate groups to create negative charges. Nitro groups and phosphate groups are used less frequently these days than they used to be because they can be used in bombmaking more readily than carboxymethyl groups. The actual charge on the functional group matters for the chromatography. In Figure 7.1, the functional groups are amines. Diethylaminoethyl (DEAE) groups are easily linked to cellulose, and are positively charged at pH values below 9. The other structure is known as "Q" and is one of many quaternary amino groups that are positively charged at all pH values.

   If a protein is going to adhere weakly to the quarternary amino group, then it has to be weakly negatively charged. It is positively charged, it will hardly stick at all, and if it is strongly negatively charged, it will adhere and never come off in any reasonable condition. It is the use of the isoelectric point (pI) of the protein that allows the lab worker to make the protein be weakly negatively charged. The pI is the average of the two median ionizable side chains on the surface. Suppose that a protein has on its surface ten aspartates (pKa=3.9), five glutamates (pKa=4.1), five histidines (pKa=6.0), five lysines (pKa=10.5), and five arginines (pKa=12.5). There is also an amino terminus and a carboxy terminus. At the median of these residues is one glutamate and one histidine (15 with pKa higher and 15 with pKa lower than these). The pI would be the average of their pKa values,

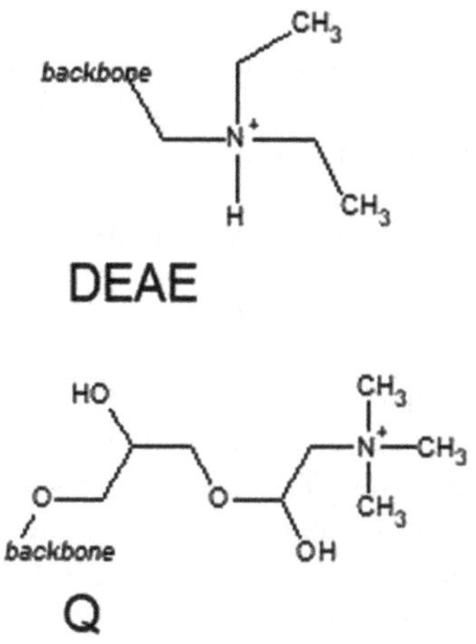

FIGURE 7.1 DEAE and Q are two common anion exchangers. DEAE has a positive charge at moderate to low pH values, and Q has a positive charge at all pH values, because it is a quarternary amine.

5.05. If the pH around the protein is less than 5.5, then the net charge of the protein is positive. If the pH is greater than 5.5, then the net charge is negative. Every pH unit higher or lower than 5.5 makes the protein more negative or more positive, respectively, by approximately one full electron charge. If the protein were kept in a buffer with pH 5.8, then it would be just slightly negative, and would have a tendency to adhere to DEAE cellulose as it flowed past this solid matrix. Other proteins whose pI values were above the pH of the buffer would not adhere at all, and would flow far past the protein you were trying to purify. No matter what comes next, you will want to get your protein to adhere weakly to the ion exchange resin, by having the pH just slightly higher than the pI if you are using an anion exchange resin such as DEAE, or just slightly lower than the pI if you are using a cation exchange resin.

After you have gotten the protein to adhere, getting the protein to dissociate again could be done in three ways: (a) You could just wait and let the equilibrium do its work as the mobile phase continues to flow past. This tends to be time-consuming and does not actually separate proteins well from the other thousands of proteins with similar pI values. (b) You could slowly increase (anion exchange resins) or decrease (cation exchange resins) the pH of the buffer being used in the mobile phase. This actually works extremely well, giving very sharp releases of proteins as the pH gradually increases or decreases. As the pH falls above or below any protein's isoelectric point, the protein is very suddenly released from the solid phase, since the value of the equilibrium constant governing its release grows extremely large when the protein takes the same charge as the solid matrix. Other proteins that had not yet seen the pH fall below their isoelectric points would not be released, giving plenty of time for your protein to be carried away by the mobile phase. (c) The third way is to outcompete the protein with some other ion. Imagine that the protein is adhering to DEAE cellulose, and that you started adding sodium chloride to the buffer. The negatively charged chloride anions would be added to the equilibrium for release of protein A:

$$A \bullet DEAE^+ + Cl^- \rightleftarrows A + DEAE \bullet Cl,$$

$$K = [A][DEAE \bullet Cl]/[A \bullet DEAE^+][Cl^-]$$

Le Chatelier's Principle makes it plain that as the concentration of Cl^- rises, the amount of unbound A must increase, since the equilibrium will shift towards the product side. This method gives almost as good a result as varying the pH, and is less expensive and easier to do. It also usually does not unfold your protein. If you do not require extremely good purification in a single step, then competing your protein off the solid phase using salt is often a better choice than using pH.

2. **Size exclusion chromatography or "Gel filtration chromatography"** separates proteins based upon their size primarily and their shape secondarily, as shown in Figure 7.2. This usually correlates well to their molar mass, so some researchers say that molar mass is the basis of separation, but it really is the size and shape of the protein. In this sort of chromatography, there is not a functional group covalently bound to the solid phase matrix. Instead, the solid phase is cracked with microscopic cracks and pits. If a protein is too large, on the other hand, it will not fit into the crack and will just be carried away by the mobile phase. The smaller the protein, the more easily and deeply it will get into the cracks. That, in turn, will make it flow the most slowly past the solid matrix. As with ion exchange chromatography, an equilibrium between the bound and unbound states is created. The reason why association occurs is different, however, so the

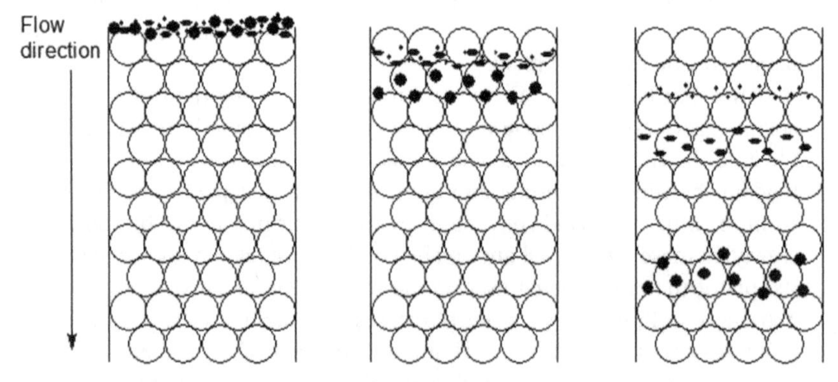

FIGURE 7.2 Gel filtration media separates proteins based upon size. Larger proteins move faster than smaller proteins when this media is used.

actual equilibrium value of K will be different for the same protein on a gel filtration matrix than it would be on an ion exchange one. This allows for a different sequence in which the same group of proteins would be eluted.

In size exclusion chromatography, the largest proteins always elute first, then medium-sized proteins, and finally small proteins are eluted last. The researcher needs to choose a solid matrix that has cracks approximately the right size for his protein. If the cracks are too small, then the protein will flow through with all the other large proteins, and will not be very pure when it elutes. If the cracks are too large, then the protein will be stuck inside the cracks in the matrix, and will not leave in any reasonable time frame. Manufacturers supply beads with cracks of defined size, which yield good separations for proteins of all different sizes. Figure 7.3 shows some of the gel filtration matrix products sold commercially. The matrix with the brand name "Superdex 200" would be effective for separation of proteins with the weight range of 10,000 Da up to about 700,000 Da. Smaller or larger than that would not be well separated for other proteins, for the reasons described. By contrast, "Toyopearl HW-75" gives effective separations for proteins with the weight range of 700,000 Da up to 500,000,000 Da. The proteins that bind to this product with be in the initial flowthrough of a column packed with Superdex 200. Choose your chromatography media after you have some idea of the size of your protein, or else the media will brutally inform you of the protein's actual size by not purifying it at all.

Sometimes, a gel filtration matrix product reveals interesting features about a protein. You might be working with a protein that has a monomer size of 30,000 Da. You might try to purify it with the Superdex 200 and discover that it exits with the flowthrough, as though it were actually very large. Instead it purifies quite well on the Toyopearl HW-75. What

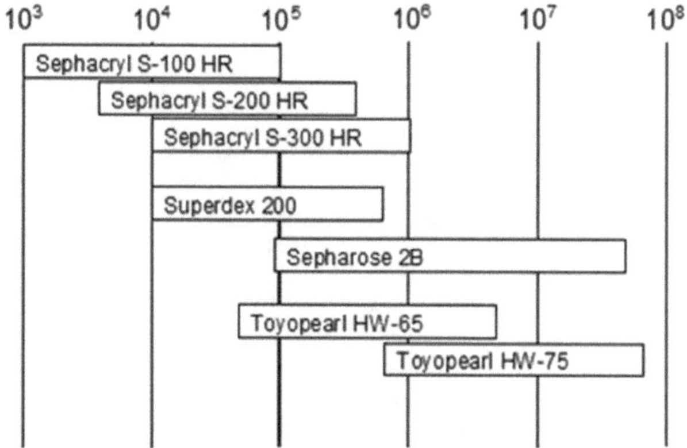

FIGURE 7.3 A number of commercial gel filtration media are available, which can separate different ranges of protein sizes. Toyopearl HW-75 would not be effective for separating proteins with molar masses between 10 kDa and 100 kDa, but it would be very effective for separating proteins with molar masses between 1 MDa and 10 MDa.

this reveals to you is that the protein monomers are joined together into an oligomer whose actual size is greater than 700,000 Da, as if 25 or 30 of the monomers are required for the protein to function. Sometimes this happens, and it is frustrating at first to "lose" your protein, only to become extremely exciting as you realize you have discovered an important structural feature.

3. **Specific Affinity** depends upon the ligand to which the protein normally binds, or some analog of it. Superficially it appears to be similar to ion exchange chromatography, but the principle behind the association of the protein and the matrix is quite different.

   Consider hemoglobin, for example. Hemoglobin biologically transports oxygen gas and carbon dioxide, but carbon monoxide and cyanide are both poisons to the function because they mimic the binding of the regular ligands. The point is that hemoglobin will adhere to them, whereas other proteins of similar size or charge might not. Even if multiple proteins both adhered to cyanide or carbon monoxide (*e.g.* hemoglobin, myoglobin, and neuroglobin) they do not necessarily do so with the same strength: they have different values of K for the equilibrium of dissociation. This is what allows something like the cyanide to be used as the principle behind binding. Cyanide groups can be covalently bound to large polymers such as cellulose, creating nitrile-cellulose. As the globins flow past this solid matrix, they will adhere specifically to the CN groups, while other proteins flow past. Then the bound globins can be released by anything that weakens their binding: pH change, salt concentration changes, mild denaturation, mild heating, competing off by adding acetonitrile or hydrocyanic acid, or a host of other treatments.

   A common ligand group that can be used for specific affinity is the Cibacron Blue dye (see Figure 7.4), variations of which mimic any nucleotide

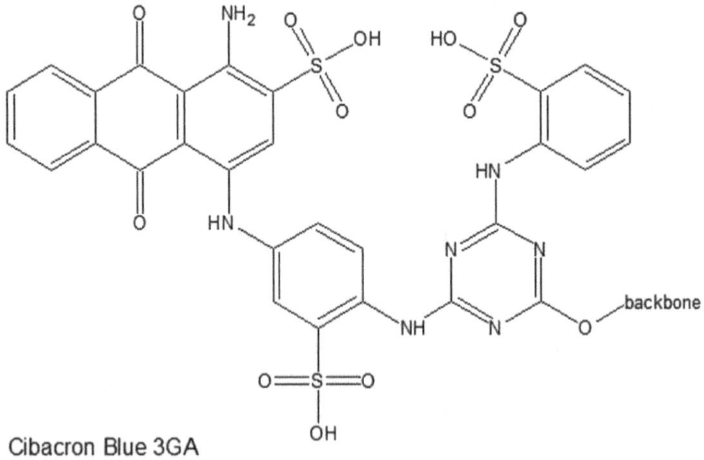

Cibacron Blue 3GA

FIGURE 7.4 Cibacron Blue is an example of a specific affinity moiety for chromatography media. Most proteins that bind nucleotides will bind to it, while proteins of similar size and charge that do not bind nucleotides will flow past it quickly.

binding protein. Things that bind ATP (the kinases) or NAD^+, FAD, or FMN (oxidoreductases) will generally have some affinity for Cibacron Blue ligands that are covalently bound to a polymer. When all other proteins that do not bind this ligand have been eluted away, the bound proteins can be eluted off by some means.

The number of possible ways to bind a protein by specific affinity is limited only by the scientist's ability to link a ligand or ligand analog to a solid matrix covalently. This is a difficult and expensive process, but almost always works out to the best possible purification. The protein can be eluted again by mass action of free ligand in a buffer solution, or else by changing ionic strength or pH in such a way that the protein's affinity for the ligand is decreased.

BACKBONE ISSUES

The functional group is only one of the considerations that goes into selecting a proper solid stationary phase matrix. The polymer backbone itself is sometimes important, too. This polymer is always going to be ground to a fine powder or a gel and the functional group that determines the type of chromatography will be formed chemically or physically. However, it is an error to think that the only chemistry that occurs is between the protein and the functional group. There is also the possibility of a chemical interaction between the polymer backbone itself and the protein.

Common biochemical backbone substances include cellulose, sepharose, dextran, sephacryl, polyacrylamide, methylacrylate, silica, alumina, and hydroxyapatite. The last three are less commonly used for protein purification than they used to be, and are not really polymers, but they do serve as mild ion exchange media. With every backbone substance, the polarity of the substance will have an effect on the purification, and that effect can enhance or diminish the quality of a chromatography purification step.

Consider what a very polar backbone (such as the cellulose in DEAE cellulose) will do to globular proteins. Most globular proteins have hydrophobic cores and surfaces with many polar residues. It is this arrangement that keeps them mostly soluble. The polar residues are the very things that interact with the functional groups in ion exchange chromatography, creating the equilibrium that separates proteins based upon their pI values.

$$\text{Protein}^- \bullet \text{DEAE}^+ \rightleftarrows \text{Protein}^- + \text{DEAE}^+, \quad \text{with its value of K}$$

where the equilibrium can be greatly shifted if the pH becomes less than the pI and the protein becomes positively charged. However, the same polar residues will have an affinity for the polar cellulose, creating a second equilibrium:

$$\text{Protein} \bullet \text{cellulose} \rightleftarrows \text{Protein} + \text{cellulose}, \quad \text{with its value of K, as well.}$$

That means that there will be two different equilibria operating at the same time with DEAE-cellulose, one undoubtedly going faster than the other. The result is that the separation of any protein from its neighbors will be somewhat smeared, especially if the protein's adherence to cellulose is measurably strong compared to its adherence to DEAE. If the value of K for the interaction with the functional group is much larger than that for the backbone, then separations are still clean. But if they are not widely different, then the solid matrix will not actually separate the proteins in your mixture. Non-polar backbone polymers such as sephacryl have less attraction or the polar residues of globular proteins and so tend to give cleaner separations due to the predominant effect of the functional group.

Of course, there are also negative features if the backbone polymer does NOT have much affinity for the polar residues. The core of most globular proteins is hydrophobic. There would be an attraction for the hydrophobic core to adhere to the hydrophobic backbone polymer, if only the protein were to unfold and expose the core. The problem is that this is exactly what does happen in the less stable proteins. If the activation energy of unfolding is somewhat low for a protein, then the presence of a hydrophobic polymer will thermodynamically stabilize the unfolded state. Thus, many proteins unfold completely when a non-polar backbone is used, instead of purifying neatly. This leads to the irrecoverable loss of the protein, rather than its purification. The polar polymer backbone might have led to less clean separations, but it also would not have destroyed the functional copies of your protein. In general, researchers prefer to use the non-polar backbone polymers if they can, due to the cleaner separations, but many times there is no real choice other than to use the more polar ones and accept that the column chromatography step will not perfectly purify the protein.

GOAL

In this lab you are going to be assigned to use one specific type of chromatography to separate L-lactate dehydrogenase and/or catalase from other proteins. You are going to compare notes with other groups to determine which technique provides the best purification (by specific activity) of your protein. You will follow only one of the methods listed, but are responsible for knowing how all of them are done.

In the interests of time and good enzyme management, you will want to assay the fractions coming off the column while they are coming off, instead of waiting for all fractions to be produced before you assay any. If you have an activity assay cocktail to make, you should prepare enough for at least 20 assays, and be prepared to make more as the experiment goes on.

If you have recovered your protein, you do not need to continue running buffer through the column. Collect two to three more fractions after your protein has passed through (or until the last fraction which has less than 10% of the total activity of what came before it). Then, you may skip past all steps that tell you to collect more fractions, and go straight to the pooling of the fractions, the analytical work, and the clean-up.

Alternative Assay for Catalase: Catalase is best tested by its activity, as has been described previously. A crude test that allows confirmatory evidence can also be found spectrophotometrically. Because catalase contains heme groups, it will absorb light strongly at 406 nm. It also tends to be brown in color. In addition to testing for activity, you should also measure the absorbance of your fractions at this wavelength. By comparing activity and absorbance, you will be able to distinguish how far catalase separates from other heme containing proteins, such as myoglobin or hemoglobin. Ask your instructor if you should perform this test for catalase, as well as the activity test.

Before starting any chromatographic technique, determine the total volume of your protein, take a 0.5 mL sample, and determine the activity, relative activity, and total activity. If asked to do so, also determine the concentration and specific activity as well. Your professor may opt not to assign determination of concentration at this time at this stage or when you pool fractions at the end of each chromatographic method.

1. IEX anion exchange chromatography, eluted by changing pH.
   a. Pack a 35 mL flash column with Q-sepharose media or another anion exchange resin by pouring a slurry of Q-sepharose in low ionic strength buffer into the column. Pour it as close to the top as you can, allowing no air bubbles to be trapped. If any air bubbles are trapped, you must empty the column and start again.
   b. Prepare 25 mL portions of 0.02 M TRIS-Glycine buffer with the following pH values: 8.0, 7.5, 7.0, 6.5, 6.0, and 5.5. You can do this by preparing 150 mL of buffer that you have adjusted to pH 8.0 with 1 M NaOH and another 150 mL of phosphate that you adjust to pH 5.5 with 1 M HCl. Then, blend the two pH buffers in the following proportions to achieve the desired pH:

Desired pH	Volume pH 5.5 Buffer (mL)	Volume pH 8.0 Buffer (mL)
8.0	0	25
7.5	5	20
7.0	10	15
6.5	15	10
6.0	20	5
5.5	25	0

   Note that blending in these proportions will only achieve the desired pH if the two source pH buffers are both the same concentration – 0.02 M in this case – and if the buffer components actually have buffering capacity this pH. Also be aware that proteins whose isoelectric point is far from the range of 7.5 to 6.0 will require a different buffer and different pH range in order to be purified by the subsequent steps.
   c. Adjust the pH of your sample of protein to 7.4 (a value somewhat above the pI of your protein) using 1 M NaOH and constant stirring. Place this sample on ice until later.

d. Set up the column so that the top and bottom are both clamped, and the bottom exit tube is much lower than the top. Attach the top to a tube that goes upwards several feet, and attach the upper tube to a reservoir of the phosphate buffer. Gravity will exert different pressure at the top and bottom of the tube, and will cause buffer to flow out of the reservoir at the top, through the column, and out the exit tube at the bottom.
e. Wash the column with approximately one column volume worth of 0.2 M Tris-Glycine buffer pH 8.0. Discard this flowthrough buffer.
f. Clamp off the top and bottom of the column to stop buffer flow.
g. Replace the buffer reservoir at the top with your protein sample. Then reopen the clamps to restart the flow, and adjust the flow so that it is going at approximately 1 mL per minute.
h. Just before the protein is completely loaded, clamp off the top and bottom of the column to stop the flow.
i. Reattach the top tube to the container with the pH 8.0 buffer. Begin the flow again. Collect approximately 5 mL fractions in test tubes. Flow 25 mL of buffer through the column, collecting fractions.
j. After 25 mL have passed through, and before any air gets into the column, stop the flow, and attach the top of the column to the pH 7.5 buffer. Restart the flow and allow 25 mL to pass through the column, collecting 5 mL fractions again.
k. Continue stopping the column flow after 25 mL, and successively flow through 25 mL of pH 7.0 buffer, pH 6.5 buffer, pH 6.0 buffer, and pH 5.5 buffer, collecting 5 mL fractions.
l. Assay the fractions for your enzyme. Pool the fractions that cumulatively contain most of the recovered activity of your protein. This does not have to be done in a rigorously analytical fashion. If you have an early peak, with the flowthrough, and a later peak, then only the later peak should be saved. The early peak is just overloaded protein, and you should have used a larger column with more solid phase material in it.
m. After you have pooled the fractions, determine the relative activity of your protein, its total volume, total activity, concentration, and specific activity. These do have to be done in a rigorously analytical fashion. Determine your percent recovery of the protein.
n. (Clean up) Unpack the column by resuspending the resin in a minimum volume of ethanol and pouring it back into the resin container from which it came. Do **NOT** pour it into the waste or into the container of any other kind of chromatography resin.

2. Size exclusion/Gel filtration chromatography
   a. Pack a 35 mL flash column with Sephacryl-300 media or some other appropriate gel filtration media by pouring a slurry of Sephacryl-300 in low ionic strength phosphate buffer into the column. Pour it as close to the top as you can, allowing no air bubbles to be trapped.

If any air bubbles are trapped, you must empty the column and start again.

b. Set up the column so that the top and bottom are both clamped, and the bottom exit tube is much lower than the top. Attach the top to a tube that goes upwards several feet, and attach the upper tube to a reservoir of the phosphate buffer. Gravity will exert different pressure at the top and bottom of the tube, and will cause buffer to flow out of the reservoir at the top, through the column, and out the exit tube at the bottom.
c. Wash the column with approximately one column volume worth of 0.02 M phosphate buffer pH 7.0. Discard this flowthrough buffer.
d. Clamp off the top and bottom of the column to stop buffer flow. Replace the buffer reservoir at the top with your protein sample. Then reopen the clamps to restart the flow, and adjust the flow so that it is going at approximately 1 mL per minute.
e. Just before the protein is completely loaded, clamp off the top and bottom of the column to stop the flow. Reattach the top tube to the buffer reservoir. Discard the first column of flowthrough, beginning from the moment you first start loading your protein sample.
f. After one column volume has passed through your column, begin collecting 3 to 5 mL fractions.
g. Assay the fractions for your enzyme. Pool the fractions that cumulatively contain 90% of the recovered activity of your protein.
h. After you have pooled the fractions, determine the relative activity of your protein, its total volume, total activity, concentration, and specific activity.
i. (Clean up) Unpack the column by resuspending the resin in a minimum volume of ethanol and pouring it back into the resin container from which it came. Do **NOT** pour it into the waste or into the container of any other kind of chromatography resin.

3. Specific affinity chromatography or Ion-exchange chromatography

The following protocol works for both types of columns, for example Cibacron Blue or Q-sepharose. A special step is indicated that optionally may be added for Cibacron Blue columns, but in many purification protocols it is not necessary. This step is in part b. In addition to adding NaCl, you may opt to add NAD+ or NADP+ or ATP, if your protein binds those ligands. Learn the value of K_d for these ligands, and to your highest salt concentration solution add enough that the ligand concentration is at ten times more than the value of K_d. The specific affinity column will release the protein because the protein preferentially binds the ligand or the salt, and the ion exchange column will release the protein because the column medium will preferentially bind the salt, due to mass action.

a. Pack a 35 mL flash column with Cibacron Blue-Agarose media or other appropriate chromatography media by pouring a slurry of the resin in low ionic strength phosphate buffer into the column. Pour it as close to the top as you can, allowing no air bubbles to be trapped. If any air bubbles are trapped, you must empty the column and start again.

b. Prepare 25 mL portions of phosphate buffer with the following NaCl concentrations: 0 M, 0.2 M, 0.4 M, 0.6 M, 0.8 M, and 1.0 M NaCl. You can do this by preparing 150 mL of 0.02 M phosphate buffer pH 7.0 with 1.0 M NaCl and another 150 mL of the 0.02 M phosphate buffer without any added NaCl. Blend the two solutions as follows in order to obtain your desired NaCl concentrations:

Desired NaCl Concentration	Volume Buffer w/o NaCl	Volume Buffer +1 M NaCl
0 M	25 mL	0 mL
0.2 M	20 mL	5 mL
0.4 M	15 mL	10 mL
0.6 M	10 mL	15 mL
0.8 M	5 mL	20 mL
1.0 M	0 mL	25 mL

In many ways, you will find this quite similar to the dilutions you would make if you were using a pH gradient on an ion-exchange column, described in Part 1.

c. Set up the column so that the top and bottom are both clamped, and the bottom exit tube is much lower than the top. Attach the top to a tube that goes upwards several feet, and attach the upper tube to a reservoir of the phosphate buffer. Gravity will exert different pressure at the top and bottom of the tube, and will cause buffer to flow out of the reservoir at the top, through the column, and out the exit tube at the bottom.
d. Wash the column with approximately one column volume worth of 0.02 M phosphate buffer pH 7.0 with no NaCl. Discard this flowthrough buffer.
e. Clamp off the top and bottom of the column to stop buffer flow.
f. Replace the buffer reservoir at the top with your protein sample. Then reopen the clamps to restart the flow, and adjust the flow so that it is going at approximately 1 mL per minute.
g. Just before the protein is completely loaded, clamp off the top and bottom of the column to stop the flow.
h. Reattach the top tube to the container with the pH 7.0 buffer without NaCl (*i.e.* 0 M NaCl). Begin the flow again. Collect approximately 5 mL fractions in test tubes. Flow 25 mL of buffer through the column, collecting fractions.
i. After 25 mL have passed through, and before any air gets into the column, stop the flow, and attach the top of the column to the buffer with 0.02 M NaCl. Restart the flow and allow 25 mL to pass through the column, collecting 5 mL fractions again.
j. Continue stopping the column flow after 20 mL, and successively flow through 20 mL of buffer+0.4 M NaCl, buffer+0.6 M NaCl,

buffer + 0.8 M NaCl, and buffer + 1.0 M NaCl, collecting approximately 5 mL fractions.

k. Assay the fractions for your enzyme. Pool the fractions that cumulatively contain 90% of the recovered activity of your protein.
l. After you have pooled the fractions, determine the relative activity of your protein, its total volume, total activity, concentration, and specific activity.
m. (Clean up) Unpack the column by resuspending the resin in a minimum volume of ethanol and pouring it back into the resin container from which it came. Do **NOT** pour it into the waste or into the container of any other kind of chromatography resin.

Follow up, after any column chromatography step:

4. Determine the total activity before the protein was purified by the column and after the protein was purified by the column. In this terminology, the "initial" state is the protein before you loaded it on the column, and the "final" state is the pooled fractions after you have eluted your protein from the column

$$\%\ \text{recovery} = \frac{\text{total activity final}}{\text{total activity initial}} * 100\%$$

5. If you determined the concentration, calculate the amount of purification you achieved

$$\text{amt.purification} = \frac{\text{specific activity final}}{\text{specific activity initial}}$$

PRE-LAB QUESTIONS

1. Which protein will be most easily purified using a Q-sepharose column: one with pI 5.5, one with pI 7.5, or one with pI 9.5, using the first method described? Explain your answer.

2. Which protein will elute first from a Q-sepharose column: one with pI 6.2, one with pI 6.5, or one with pI 6.8, using the third method described? Explain your answer.

3. Which protein will elute first from a Sephacryl-300 column: one with MW 50 kDa, one with MW 100 kDa, or one with MW 150 kDa? Explain your answer.

POST-LAB QUESTIONS

1. You are purifying an enzyme which has a relative activity of 400 U/mL, and you have 20 mL of it. You purify it by ion exchange chromatography, obtaining 5 mL fractions. Measuring the activity of the fractions, you obtain the following results:

Fraction	Relative Activity (U/mL)
1	0
2	0
3	0
4	0.1
5	2
6	400
7	360
8	250
9	150
10	100
11	70
12	50
13	35
14	21
15	9
16	2
17	0.1
18	0
19	0
20	0

a. What is the initial total activity of your enzyme?

b. If you pooled all your fractions, what would be your percent recovery?

c. Which fractions should you collect in order to recover at least 75% of your original activity with the greatest possible purity?

d. Explain what might cause the percent recovery in Part b to be less than 100%.

e. It is better to omit fractions 4 and 5 when pooling to obtain 75% of the activity, as done in Part c. Explain why.

2. Suppose you initially have 250 U of protein activity. You load the protein onto a Q-sepharose column. In a peak that comes immediately after loading, you find 30 U of activity, and in five fractions later you pool together a total of 110 U. Explain (a) why you have the two peaks of activity, and (b) why there are only 140 U that you can account for at the end.

8 Michaelis–Menten Kinetics

Enzymes are biological catalysts, and generally proteins. They enhance the speed and specificity of reactions to such an immense degree that many people think of them as "magical". The use of this term connotes a tendency to think that they will violate or circumvent normal natural laws. Intellectually, scientists all admit that living systems must obey the same natural laws as the rest of the world. However, as a purely emotional response and often without thinking, enzymes are treated as if they were not performing ordinary chemical catalysis.

Enzymes are ordinary chemical catalysts. They obey the same rules, use the same kinds of mechanisms, and are described by the same mathematical formulations as non-enzyme catalysts. They just tend to be more efficient.

All catalysts bind a substrate, then provide a mechanism for that substrate to react to a product as shown by the scheme below:

Step 1, fast:	$E+A \rightleftarrows EA$	$Rate_1 = k_1[E][A] = Rate_{-1} = k_{-1}[EA]$, $K_A = [E][A]/[EA]$
Step 2, slow:	$EA \rightarrow EP$	$Rate_2 = k_{cat}[EA]$
Step 3, fast:	$EP \rightarrow E+P$	$Rate_3 = k_3[EP]$

In this scheme, E is the catalyst, A is the substrate, and P is the product. k_1, k_{-1}, k_{cat}, and k_3 are the rate constants for the steps, with "$_{-1}$" designating that it is the reverse of step 1. K_A is the equilibrium constant for the dissociation of A from the EA complex, and is also the ratio k_{-1}/k_1. Since step 3 is so much faster than step 2, the rate of the process is dictated by the rate of step 2, and it can all be simplified to the following scheme:

$$E + A \rightleftarrows EA \rightarrow E + P$$

This is the schematic worked out by Leonor Michaelis (1875–1949) and Maud Menten (1879–1960).

The equilibrium constant K_A is replaced by the "Michaelis" constant K_m, which measures both the affinity of the ligand, or the "substrate", for the catalyst and the relative proportion of loss of the EA complex to simple dissociation or to catalysis. With many enzymes, it is numerically equal to K_A, and only differs in meaning if catalysis is extremely fast compared to dissociation. That isn't often the case, though it does happen sometimes if the rate constant k_{cat} is not insignificant compared to k_{-1}.

In all cases, the rate itself will depend upon the percentage of enzyme that is in the complexed state and the magnitude of k_{cat}. As the amount of substrate is increased, there will be a larger percentage of enzyme that is complexed, and so the rate increases. This is not a linear relationship between rate and substrate concentration, but rather is a rectangular hyperbolic relationship, depending upon the concentration of substrate and the value of K_m.

$$\text{Rate} = v_0 = V_{max}[A]/(K_m + [A])$$

where v_0 is the rate before any significant product is formed, V_{max} is the product of k_{cat} and the total enzyme concentration, [A] is the substrate concentration, and K_m is the affinity of the substrate for the enzyme, as described earlier.

Suppose you are studying the activity of catalase that you have extracted from yeast. As we have seen, the assay merely involves adding an amount of catalase to a solution of hydrogen peroxide, then measuring the rate of reaction by timing how long it takes for a certain amount of oxygen gas to form, accounting for the dilution of the enzyme as well. If you were to raise or lower the concentration of the hydrogen peroxide, you would see the rate increase or decrease. In this hypothetical case, you get the following rates when you vary the concentration of hydrogen peroxide:

H_2O_2 (mM)	Rate (μmol/min)
0.3	6000
0.6	12000
1.5	23000
2.9	32000
5.9	43000
15	56000
29	63000
59	64000
150	67000
290a	67000
590	68000
880	68000

When we plot these data on a regular plot, seen in Figure 8.1, the classic Michaelis–Menten curve is revealed, clearly indicating at least the value of V_{max}. However, there will be occasions when the value of K_m will not be as clear, as is the case this time. The curvature is so steep that it is hard to identify which concentration gives a rate which is ½ V_{max}.

There are a couple of good possibilities in such cases. One that is easily available with modern computers is to replot the data in a "semilog" plot, in which equal distances refer to increases on the power of ten on the x-axis, but equal distances are directly proportional to the value of the data on the y-axis. A semilog plot can make it harder to visually identify V_{max} but easier to identify K_m. The concentration that equals K_m will occur at the inflection point of the curve of best fit on a semilog plot, as seen in Figure 8.2.

A good graphing program capable of linear and non-linear regression will be able to solve for K_m and V_{max} without needing to plot it as a semilog plot, but humans determining these values by eye often will prefer the semilog plot for finding K_m and

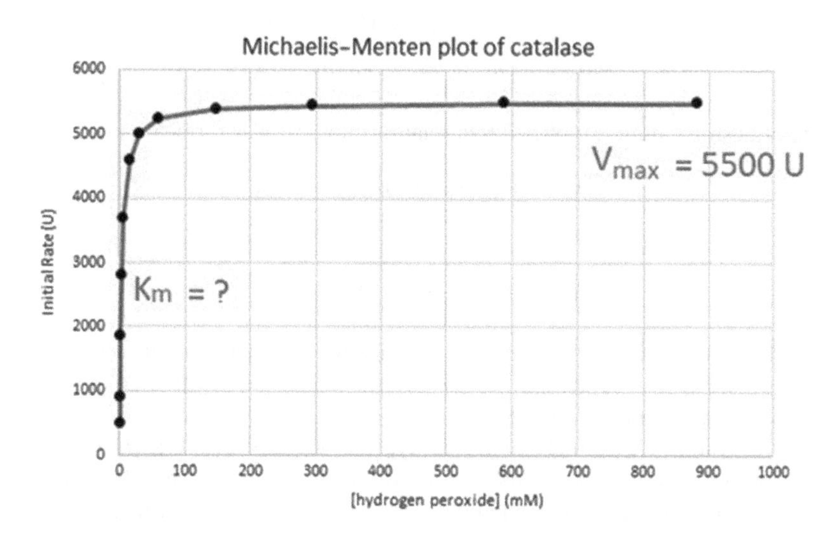

FIGURE 8.1 The Michaelis–Menten plot of enzyme activity, plotted on a proportional x-axis, reveals the maximal activity with clarity, but it is harder to discern the value of the Michaelis constant.

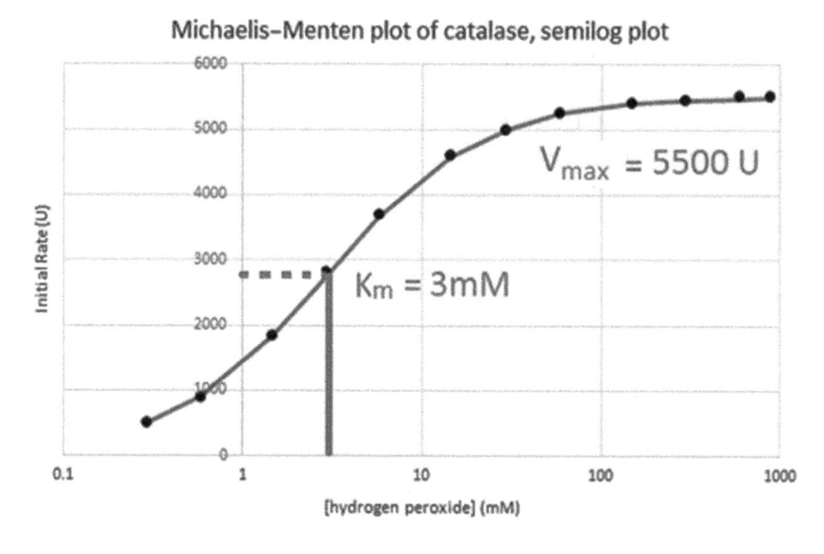

FIGURE 8.2 When the Michaelis–Menten plot is created using a semilog x-axis, it is much easier to determine the value of the Michaelis constant, so long are there are sufficient data points on either side of the half-maximal activity value.

the proportional plot for finding V_{max}, as these parameters are exaggerated in each respective visualization.

There is another approach, which has its own strengths and weaknesses. An easy way to determine V_{max} and K_m is by converting the non-linear Michaelis–Menten formulation of rate into the linear reciprocal of rate.

$$\text{Rate} = v_0 = V_{max}[A]/(K_m + [A])$$

$$1/\text{Rate} = 1/v_0 = (K_m + [A])/V_{max}[A]$$

This relationship can be simplified to a relationship of linear form (y=mx+b):

$$1/v_0 = (K_m / V_{max}) * 1/[A] + 1/V_{max}$$

where

$$m = (K_m/V_{max}) \quad \text{and} \quad b = 1/V_{max}.$$

Now, if the reciprocal of initial rate is plotted against the reciprocal of concentration, both V_{max} and K_m may be determined by the y-intercept and by the ratio of the slope to the y-intercept, seen in Figure 8.3.

If the data are of extremely high quality, then K_m will also equal the negative x-intercept, but determining it this way usually leads to large errors. It is not common to get data that are reliable enough to be certain the apparent x-intercept is correct, among other reasons.

This sort of plot is referred to as the "Lineweaver–Burke" plot, for the biochemists who developed the method of solving kinetic parameters this way. It is extremely easy to do and usually requires nothing other than a straight edge and some graph paper. On the other hand, it has a serious flaw, which can be seen in the graph itself. Data points that come from low concentrations (those on the right) are given more weight than data points from high concentrations (on the left) because they are more spread out. However, the actual rates determined at lower concentrations have higher errors associated with them, and so they ought to be given LESS

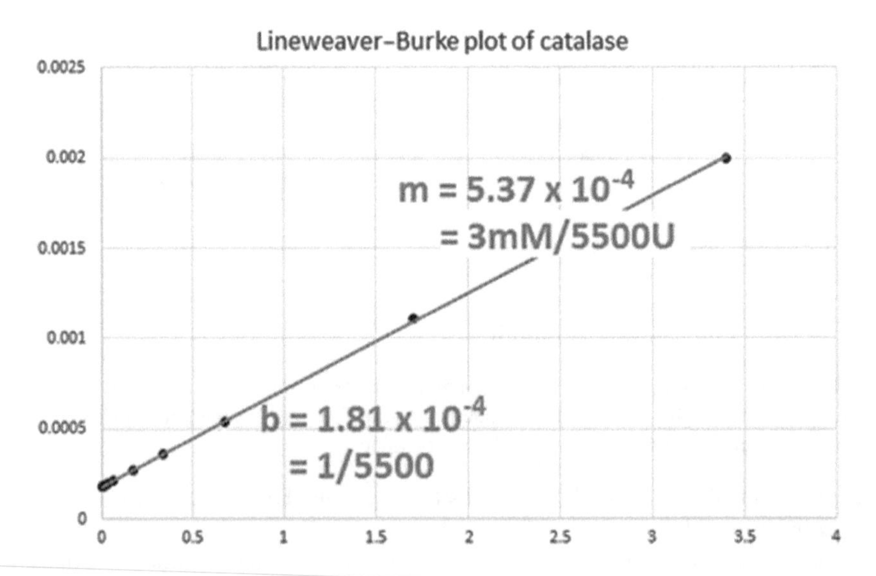

FIGURE 8.3 A Lineweaver–Burke plot can be used to determine both the maximal activity and the Michaelis constant.

weight. This means that the slope is usually affected very strongly by unreliable data points, skewing the value of K_m more than it should be. As mentioned in the previous paragraph, this isn't a problem if the error associated with the lower data points turns out to be very low, but it is an unreliable assumption to think that low error would come with these data points. If there is a disagreement between the value of K_m solved by a semilog Michaelis–Menten plot or by non-linear regression and the value as solved by a Lineweaver–Burke plot, then the latter is usually not to be trusted.

When V_{max} and K_m have been determined with some certainty, it becomes possible to deduce both the catalytic rate constant, k_{cat} and also the catalytic proficiency, k_{cat}/K_m. Because V_{max} is defined as:

$$V_{max} = k_{cat}[E]_T,$$

We can see by simple rearrangement that:

$$k_{cat} = V_{max}/[E]_T$$

where the total enzyme concentration, $[E]_T$ is measured in molar terms. Generally, the total enzyme concentration is known in units of "mg/mL", but if they molar mass of the enzyme is known, then it can be converted to moles per liter, or "M", quite easily.

For example, consider an enzyme whose concentration is 7.2 mg/mL with a molar mass of 410 kDa. This enzyme is showing V_{max} equal to 500 units per minute of activity and a K_m of 0.13 M. With this information, it is a simple matter to calculate the total enzyme concentration in molar terms and subsequently the rate constant:

$$[E]_T = 7.8\frac{\text{mg}}{\text{mL}} * \frac{1\text{ g}}{1000\text{ mg}} * \frac{1000\text{ mL}}{\text{L}} * \frac{\text{mole}}{410{,}000\text{ g}} = 1.9 * 10^{-5}\text{ M}$$

and

$$k_{cat} = \frac{500}{\text{min}} * \frac{\text{min}}{60\text{ s}} * \frac{1}{(1.9 * 10^{-5}\text{ M})} = 4.4 * 10^{5}\text{ M}^{-1}\text{ s}^{-1}$$

When we know the value of the catalytic rate constant, we can combine it with the value of the Michaelis constant to find out the catalytic proficiency, sometimes considered the "specificity" of the enzyme for catalyzing a particular substrate:

$$\frac{k_{cat}}{K_m} = \frac{4.4 * 10^{5}\text{ M}^{-1}\text{ s}^{-1}}{0.13\text{ M}} = 3.4 * 10^{6}\text{ M}^{-2}\text{ s}^{-1}$$

COOPERATIVITY

The majority of enzymes have multiple subunits, and multiple active sites that function at the same time. When this happens, it is possible that the activity occurring at one site will influence the activity occurring at the other active sites. Sometimes, catalysis at one site makes it easier for catalysis to occur at another site in a phenomenon called "positive cooperativity". Sometimes, the reverse is true, and catalysis at one site inhibits any activity at other sites. This phenomenon is called "negative cooperativity" and it is responsible for the observation of some enzymes being only "half-the-sites-active". In both cases, the cooperativity can be modeled with high-quality data fitted to a Michaelis–Menten plot, which has been slightly modified:

$$\text{Rate} = v_0 = V_{max}[A]^{nH}/(K_m^{nH}+[A]^{nH})$$

The exponent "nH" is known as the "Hill coefficient". Its value ranges from "0" at a minimum up to the number of active sites at a maximum. When it equals "1", this equation reverts to the regular Michaelis–Menten equation, and it indicates that no cooperativity is occurring. When the Hill coefficient is greater than 1, positive cooperativity is occurring. A sigmoidal shape to the best fit of the data seen on a proportional plot is characteristic of positive cooperativity, shown in Figure 8.4.

Catalase from most species does not exhibit cooperative behavior, but many enzymes do. Likewise, many enzymes have more than one substrate, which can induce a phenomenon much like cooperativity. K_m for one substrate may change at different concentrations of another substrate. One substrate may bind without cooperativity in the absence of the second substrate, but with increasingly cooperative behavior in its presence. Such things can be found by kinetic analysis, and they will indicate whether "intramolecular communication" is occurring with your enzyme.

The enzyme selected for this laboratory has only one substrate and should not display cooperative behavior, but expect to find both positive and negative

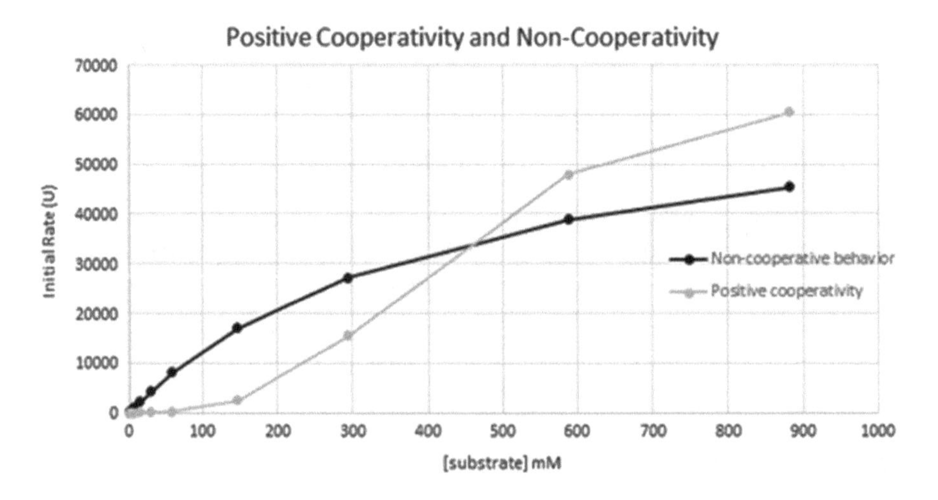

FIGURE 8.4 Positive cooperativity will result in a sigmoidal shape to the line of best fit on a Michaelis–Menten plot whose x-axis is proportional. This is not to be confused with the apparent sigmoidicity of the fit when the data is plotted on a semilog x-axis.

cooperativity as you characterize other enzymes throughout your career. It is quite common.

DATA IN TRIPLICATE

As a final note, it is better to take all data points in triplicate and determine average initial rate as you vary the concentration. There can be considerable variation from the expected Michaelis–Menten kinetic pattern if you only take one measurement of rate, but this will only be because of the error in your individual measurement. When you take the average value of multiple data points, the error is greatly reduced.

For example, if at 0.01% hydrogen peroxide, you find you need a 1:500 dilution of 8 mg/mL enzyme to give activities of 26000 U, 52000 U, and 18000 U, you are more likely to converge upon an accurate value of activity if you accept the average:

$$(26000\ \mathrm{U} + 52000\ \mathrm{U} + 18000\ \mathrm{U})/3 = 32000\ \mathrm{U}$$

32000 U will work better with the average values you take at other concentrations of hydrogen peroxide than any of the individual values does, as seen in Figure 8.5.

GOAL

The purpose of this experiment is to kinetically characterize the enzyme catalase, determining the value of K_m, V_{max}, and k_{cat}.

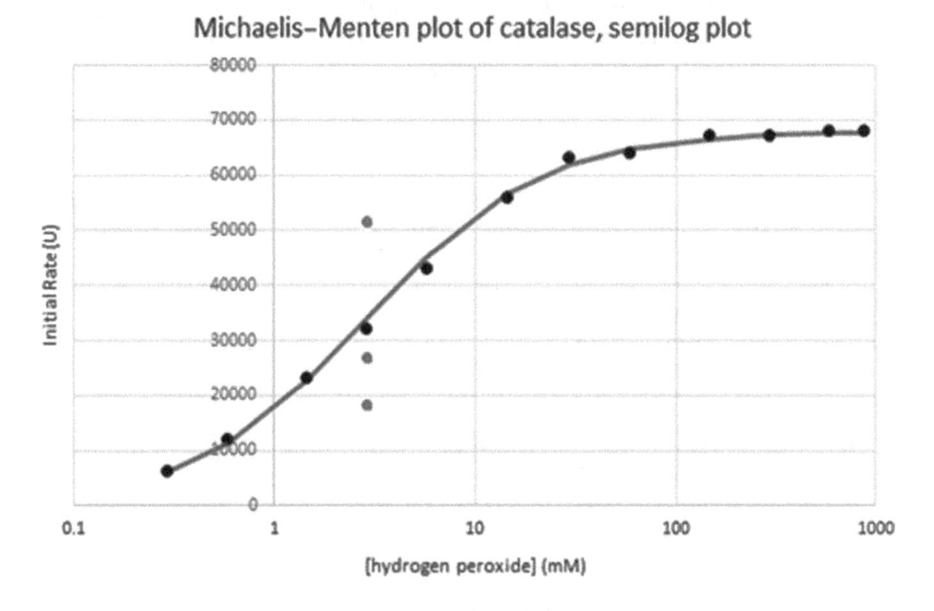

FIGURE 8.5 A line of best fit is greatly improved when multiple measurements are taken at the same concentration, and then averaged together. None of the individual trials taken at 5mM are on the line of best fit, but the average of those trials is closely modeled by that line.

Procedure

Catalase: A Simple Assay

The assay procedure will be a variation on what you have tried before. If one is merely determining the presence of the enzyme catalase, then having the enzyme work with maximal speed is best. One would, in such a case, use enough hydrogen peroxide to guarantee that the enzyme will be concentration insensitive to the substrate. The assay is simply to put 5 μL of the enzyme solution on a standard hole-punch-size paper disk of Whatman #2 paper, drop the enzyme-saturated disk into 10 mL of the hydrogen peroxide, and record the amount of time that elapses until the disk reaches the surface. The volume of oxygen gas produced is approximately 10 μL, as described previously. If the temperature and pressure of the room are independently measured, then the moles of oxygen gas can be determined, which reveals the micromoles of reaction catalyzed within the time it took for the disk to rise, and thus the activity of the enzyme is revealed. However, this time, vary the procedure according to the following steps:

1. You will need some 50 mL beakers, a 10 mL graduated cylinder or a 10 mL transfer pipet, a stock of 3% hydrogen peroxide, some 0.02 M phosphate buffer pH 7.0, a stopwatch, some forceps or tweezers, a number of paper disks of Whatman 40 filter paper or Whatman #1 filter paper punched out by a standard notebook hole-puncher, and a 10 μL adjustable pipettor. You will also need to know the temperature and barometric pressure.
2. Your H_2O_2 concentrations should be 0.001%, 0.002%, 0.005%, 0.01%, 0.02%, 0.05%, 0.1%, 0.2%, 0.5%, 1%, 2%, and 3%. Prepare them in *different* beakers than the ones in which you are going to assay the catalase, by serial dilution, as shown in the table below:

Desired Conc (%w/v)	Source	Volume Source (mL)	Volume Buffer (mL)
3%	3%	90	0
2%	3%	60	30
1%	3%	30	60
0.5%	3%	15	75
0.2%	2%	9	81
0.1%	1%	9	81
0.05%	0.5%	9	81
0.02%	0.2%	9	81
0.01%	0.1%	9	81
0.005%	0.05%	9	81
0.002%	0.02%	9	81
0.001%	0.01%	9	81

Please note that if you are limited on the number of beakers you have, you may need to make some of the solutions on this list, test them as described in the next steps, clean half of your beakers, and make the next dilutions in them. Start with the highest concentration and work downwards.

3. Record the temperature and pressure.
4. Obtain an Eppendorf tube with some catalase with a concentration of 1 mg/mL in phosphate buffer pH 7.0. Keep this on ice. In further Eppendorf tubes, prepare a 1:10 dilution of this enzyme, a 1:50 dilution, a 1:100 dilution, a 1:500 dilution, a 1:1000 dilution, and a 1:5000 dilution. Keep all these dilutions on ice.
5. Dispense 10 mL of 3% H_2O_2 into a 50 mL beaker. Pick up a paper disk with the forceps. With the pipettor, add 5.00 μL of the 1:1000 dilution of catalase to the paper. Drop the paper into the hydrogen peroxide, and start the stopwatch right away. The paper should sink immediately, and bubbles of oxygen will start to form and cling to the paper. When the density of the paper is less than that of water, it will float to the surface. Stop the time at that point. Record your data, then repeat this measurement twice, so that you have data in triplicate.
   a. If the time it takes for the disk to rise to the surface is less than 10 seconds, repeat Step 4 with the 1:5000 dilution, and if necessary make a 1:10000 dilution as well. If the time it takes for the disk to rise is more than 2 minutes, then repeat step 4 with a less dilute solution of catalase.
6. Repeat the previous step with each concentration of H_2O_2, going downward in concentration. If at any point it becomes apparent that you must use a new dilution of catalase in order to get a response within 10 seconds to 2 minutes, then repeat your measurement with the new concentration. Measure the activity for all concentrations of H_2O_2.
7. Calculate all your H_2O_2 concentrations in "M" using the molar mass of hydrogen peroxide (MW=34.02 g/mole) and the fact that "%" means "grams per 100 mL of solution".
8. Determine the average activity at each concentration of your three measurements.
9. Plot the average activity versus the concentration of hydrogen peroxide. Determine V_{max} and K_m visually, if possible.
10. Determine the values of $1/[H_2O_2]$ and $1/V_0$. Make a Lineweaver–Burke plot of these data. Determine the best linear fit. Calculate V_{max} and K_m.

$$V_{max} = 1/\text{y-intercept}$$

$$K_m = \text{slope/y-intercept}$$

11. Compare your values as determined by the two methods. Evaluate which one is more reliable, given the quality of your data set. The specific activity of catalase from *Aspergillus niger* is approximately 4500 U/mg and K_m is approximately 0.3 M.
12. (Bonus) Using the enzyme concentration, determine the value of k_{cat} for this enzyme, since V_{max} equals $k_{cat}[E]_{total}$.

PRE-LAB QUESTIONS

1. You need to make 10 mL volumes of H_2O_2 with concentrations of 2%, 1%, 0.5%, 0.2, 0.1%, and 0.05%. You have the usual stock bottle of 3% H_2O_2. List the volumes of stock and of distilled water that you will use to make these dilutions.

	Vol Stock	Vol H_2O
2%		
1%		
0.5%		
0.2%		
0.1%		
0.05%		

2. In the preceding question, should you actually use the 3% stock to make all the dilutions? If not, for which concentrations would you use a different stock, and what concentration should that different stock be?

3. If you measure the activity of a 1 mg/mL enzyme at different concentrations of substrate and obtain the data listed below, determine the values of K_m, V_{max}, and k_{cat} for your enzyme.

[substrate] (mM)	Activity (μmole/min)	1/[substrate]	1/Activity
0.1	222		
0.2	364		
0.3	462		
0.8	696		
2	851		
5	935		
10	966		

You will need to fill out the rest of the chart and conduct Lineweaver–Burke analysis to complete this problem.

POST-LAB QUESTIONS

1. Mathematically prove from the Michaelis–Menten rate equation that Km = [S] when the rate v = 1/2 Vmax.

2. Consider the Michaelis–Menten scheme $E + A \leftrightarrow EA \rightarrow E + P$. Under what circumstances will the rate law be (i) pseudo-zero order with respect to substrate; (ii) pseudo-zero order with respect to enzyme; (iii) first order with respect to both substrate and enzyme?

3. What are the inherent weaknesses of the Lineweaver–Burke plot? Assume you have 5% error in all of your determinations of rate: What is the error of 1/rate at your highest and lowest H2O2 concentrations?

9 Protein Purification

A protein purification is not something that can be fit into a couple of hours of work. Instead, it involves tense hours of labor, followed by tedious hours of waiting. The reward at the end, however, is very great for researchers: proteins can be understood best when they are isolated. You can characterize everything from the complete activity of every isozyme down to the structure of the molecule.

The downside is that your life will rotate around your need to meet a timetable during a purification, not the other way around. From the moment that you first disrupt the source tissue for your protein, it will begin decomposing. At the same time you disrupt a cell to get at a protein, you are releasing it and also all the proteases that are in the cell. They will begin to hydrolyze every protein they can, including the one you are trying to purify. You have to get it away from the proteases as quickly as you can. You also have the problem of oxidation due to exposure to the atmosphere, whereas the insides of cells are almost always reducing. Your protein will have groups such as cysteine readily oxidize, changing the isoelectric point and possibly forming unwanted disulfide bonds. Other oxidization possibilities exist, all of which lead to loss of structure and loss of function of the protein. The point is that the longer your purification takes, the less product you will have.

There are many occasions where a scientist doing a protein purification can expect to start work at sunrise on one day and will not finish until sunrise the next day. In professional situations, teamwork is required in order to have success in situations like this, because it is vital to maximize the amount of pure product you recover. In an educational lab, it is more common to accept that you will not recover the highest possible amount of project. Generally, in laboratory class situations, you could expect to work in three- to five-hour blocks during a week, and then not resume until a week later. The recovery from one week to the next will be lower in most cases, but you accept that penalty as the price that must be paid to meet all other academic obligations while still learning the technique.

There is very little new to the "technique" that has not been performed in previous chapters of this book. Purification calls for some combination of the skills of salting out and centrifugation, as well as various kinds of chromatography, constantly blended with determinations of concentration, and assays to make sure you know where your active protein is at all times and how much you have. As can be imagined, the real skill that is learned in doing a purification for the first time is managing your time such that you do all these tasks simultaneously and efficiently.

DIALYSIS

One technique that will be introduced in this session is *dialysis*. It will be used as the stopping point in each block of work that must be done.

Dialysis is the exchange of one ion background in a small volume for the ion background in a larger volume. It is a controlled form of dilution. The solution containing your protein and whatever buffer and salt condition the protein is presently within is collected, then put into a dialysis membrane of some kind. A dialysis membrane often takes the form of a small bag that is tied off to prevent leakage. The material in this bag is "semipermeable" which means that water can flow across it along with all small ions and molecules, through small pores that are in the material. On the other hand, molecules the size of proteins are too large to cross through the material. Therefore, salt and buffer components will flow in and out of a dialysis membrane, but the protein must remain in whatever space it was placed within at the start. This results in the change of the ion and buffer background to the protein.

Consider the following example. You have recovered your protein in a 40% saturated ammonium sulfate supernatant fraction in 0.02 M phosphate buffer at pH 7.0. The volume is 20 mL. You put it in a dialysis bag which you tie off, then place the bag into 975 mL of 0.02 M phosphate buffer at pH 7.0. What you want is to get the ammonium sulfate salt to dialyze away. What would the concentration be, after the salt has reached an equilibrium of dialysis? The combined volume is now 995 mL, so the new concentration can be found from knowing this combined volume, the previous volume, and the previous concentration:

$$\text{Conc of } (NH_4)_2SO_4 = (20 \text{ mL}/995 \text{ mL}) * 40\% \text{ saturation} = 0.8\% \text{ saturation}$$

This is not quite zero ammonium sulfate, but it is obviously much lower than it was before. If one did multiple rounds of dialysis, then within significant figures, the value approaches zero within two or three rounds at the most. Functionally, that is almost always what is required.

If you are using a dialysis bag, the usual procedure is to cut a length of the back that is about 40% more than the volume you will need to hold your protein. Wet it thoroughly on the inside and out with deionized water or buffer. Tie off one end, and check to make sure the bag is not leaking. If it is secure, then pour your protein into the bag and tie it off at the top. Make sure to leave some room for the inevitable expansion of volume, due to the osmotic pressure from the protein itself. Place the bag into a reservoir of buffer, between 500 to 1000 mL in volume. Generally, this is done with a large beaker. After the bag is in the beaker there is an extra optional step, if you have magnetic stir plates in a cold room to use: add a stir bar to the beaker. Whether you add the stir bar or not, cover the beaker with parafilm, and place it in a cold room or other refrigerated space. Dialysis will be rapid with stirring (six to twelve hours), but will occur completely over two to five days even without stirring.

PURIFICATION TABLES

When you purify a protein, you need to keep track of several things: (1) where your protein is; (2) how much of your protein is active; (3) how pure your protein is. Doing so will allow you to evaluate how successful a particular purification step is. If you find that a purification step loses a lot of active protein, or makes a protein less pure than it was before the step was performed, then you would consider not repeating that purification step in future work. You must document your information at each stage, if you are going to be able to make this kind of evaluation. For this reason, you keep a purification table.

A purification table names the step you perform, and records the volume at that step, the activity, the relative activity, the total units of activity, the concentration, the specific activity, the purity, and the step efficiency. It also is helpful to indicate whether the protein is found to be present in a particular step.

Name	Volume (mL)	Activity (U)	Rel. Act (U/mL)	Total Act. (U)	Conc. (mg/mL)	Spc. Act. (U/mg)	Purity Factor	Efficiency
Crude*	13.2	192.46	38492	508000	40	962.3	N/A	N/A
40% pel	20.0	8.025	1605	32100	9.0	N/A	N/A	N/A
40% sup*	11.9	192.02	38403	457000	30	1280	1.33	1.20
65% pel*	18.0	101.66	20333	366000	9.5	2140	2.22	1.33
65% sup	21.5	9.558	1912	41100	8.3	N/A	N/A	N/A
1st dial*	24.1	51.245	10249	247000	6.4	1601	1.66	0.50
Q-seph*	27.7	35.74	7148	198000	4.5	1588	1.65	0.80
2nd dial*	28.0	14.11	2821	79000	1.8	1567	1.62	0.39
Blue*	16.6	20.24	4048	67200	2.6	1557	1.62	0.85
3rd dial*	18.4	3.968	793.5	14600	0.51	1556	1.62	0.22

For names, you use a short description of the step, which you make sure you indicate within your notebook, for fuller information. You also indicate which fractions were used for the further steps by some sort of mark, an asterisk in this case (*). Invariably, these fractions are used because they have the most units of activity of your protein or else they represent the greatest level of purity. For example, "crude" refers in this case to the soluble protein portion from tissue that has been disrupted by sonication and separated from insoluble materials via centrifugation. Your notebook would clearly define the meaning of your name for a step, as a "legend" for your table. As an example, the meaning of the names of the sample table is shown below:

Crude: the soluble protein portion from tissue that has been disrupted by sonication and separated from insoluble materials via centrifugation.

40% pel: the pellet that formed upon addition of ammonium sulfate to a 40% saturation level, separated from the supernatant by centrifugation, and resuspended in 20 mL of 0.02 M phosphate buffer, pH 7.0.

40% sup: the supernatant that remained after addition of ammonium sulfate to a 40% saturation level, separated from the pellet by centrifugation.

65% pel: the pellet that formed upon addition of ammonium sulfate to a 65% saturation level, separated from the supernatant by centrifugation, and resuspended in 18 mL of 0.02 M phosphate buffer, pH 7.0.

65% sup: the supernatant that remained after addition of ammonium sulfate to a 65% saturation level, separated from the pellet by centrifugation.

1st dial: the protein recovered from the 65% pellet which was dialyzed against 0.02 M phosphate buffer, pH 7.0.

Q-seph: the combined fractions exhibiting the protein's enzymatic activity, which eluted from a Q-sepharose column at decreasing pH values, representing 90% of the total recovered activity.

2nd dial: the protein recovered from the Q-seph fractions which were dialyzed against 0.02 M phosphate buffer, pH 7.0.

Blue: the combined fractions exhibiting the protein's enzymatic activity, which eluted from a FastFlow Blue-sepharose column at increasing NaCl concentrations, representing 90% of the total recovered activity.

3rd dial: the protein recovered from the Blue-seph fractions which were dialyzed against 0.02 M phosphate buffer, pH 7.0. The final product.

This chart represents both an accounting of what you did and the plan of what you are going to do in order to purify your protein. You populate the table as you complete each step. The numerical values on the table should take many hours of work to complete. Also, you sometimes leave blank spaces for material you know you will discern as you go along. In this case, the term "65% pellet" is underlined in the description of the **1st dial** entry, because until the protein was recovered, there was no way to know that it was going to be the 65% pellet and not some other ammonium sulfate fraction which would wind up being dialyzed.

To determine the **purity factor**, divide the specific activity of a particular step by the specific activity of the first step (the "crude" in this case). If the value increases, the purity has increased by that factor. This is because the specific activity is a measure of what proportion of your total protein is the active protein you seek, *i.e.* its purity. For example, the 65% pel step displayed a specific activity of 2140 U/mg, whereas the crude displayed only 962.3 U/mg. The purity factor is therefore:

$$\text{Purity} = 2140/962.3 = 2.22$$

This number indicates that the protein is 2.22 times more pure at the completion of this step than it was at the start. When one compares it to the preceding step, the 40% sup, which had a purity factor of 1.33, we can see that the protein became more pure as a result of subjecting it to the 65% ammonium sulfate cut. However, as one examines the table for this purification, it is clear that the next step containing the protein, the 1st dial, has a lower specific activity, and a purity factor of only 1.66. This means that exposing it to dialysis actually harmed the purity of the protein. Subsequent steps in the sample purification don't improve from a purity of 1.66 and actually decrease slightly. These numbers indicate that these final steps did not improve the purity of the protein at all.

The **efficiency** term allows you to evaluate how valuable a purification step you had, given how many active units of protein you were able to recover and how much purity you achieved at the cost of sacrificing some of the active units. You never gain in active units of protein, in real life, so some loss is to be expected. The scientist must determine if the cost of sacrificing those active units was worth it in return for some gain in purity. This efficiency term allows such a determination to be made. It is calculated as the ratio of the product of total activity and specific activity of one step divided by the product of those terms from the previous step. If this ratio is higher than one, then the performed step gave more purity than it cost in activity, and is worth retaining.

For example, compare first the 40% sup to the crude step. The efficiency is:

$$\text{efficiency} = \frac{457000 * 1280}{508000 * 962.3} = 1.20$$

The efficiency value of 1.20 indicates that this was a productive step.

On the other hand, compare the 2nd dial step to the Q-seph step which preceded it. The efficiency is:

$$\text{efficiency} = \frac{79000 * 1567}{198000 * 1588} = 0.39$$

With an efficiency value of only 0.39, it is plain that the step sacrificed almost three times as much in recoverable protein as it might have, and in return got no increase in purity at all. This was a terrible step, and should not have been performed the way it was performed. The purification documented in the sample table is clearly a failure, with multiple steps that resulted in worse purification than would have occurred if those steps had never been done. You should attempt to develop purifications in which every step has an efficiency higher than 1.

GOAL

The purpose of this experiment is to purify either catalase or L-lactate dehydrogenase from a sample of beef liver, maximizing both the purity of the product and the total recovery.

METHOD

For your protein purification, you will combine the methods you have thus far learned. You will do them on different days, punctuated by *dialysis*. Instead of giving the specific steps, the overall strategy will be to repeat the experiments as you have done them, but on a specific timeline and in a specific sequence.

Be certain to set up a purification table, as described above, and update it every lab session. Be certain to show all calculations you perform in order to fill out the table.

First Lab Session

Extract your protein from the source. 5 g of beef liver will be homogenized in 80 mL of 0.02 M phosphate buffer at pH 7.0. You will be given approximately 20 mL of it. Then follow the procedure to "salt out" your protein with ammonium sulfate. At the end, dialyze your protein against the phosphate buffer, in a volume between 500 mL to 1 liter.

At each stage, save a 0.5 mL sample in a labelled Eppendorf tube. Keep these samples on ice. Determine the activity and concentration of all your samples. Activity must be determined very soon after the sample is taken. Concentration for all saved samples may be done either on the first lab day after the enzyme is placed in dialysis, or else all samples can be frozen and the concentration is determined in the third week.

Second Lab Session

Remove your protein from dialysis and determine how much activity remains. Then purify the protein on a chromatography column, usually size exclusion or ion exchange. Pool the fractions that give the highest amount of recovery of your most active chromatography peak. Dialyze them against phosphate buffer at the end, just as you did in the previous session, if your method changed the ion background at all, or if you will require it to be in a different ion background for the next session. Otherwise, just store it safely in a refrigerated space until you are ready to use it again.

As before, at each stage, save a 0.5 mL sample in a labelled Eppendorf tube. Determine the activity and the concentration of each sample at the appropriate time.

Third Lab Session

Remove your protein from dialysis or storage and determine how much activity remains. Then purify the protein using a second method of chromatography, one that you have not used in this purification strategy and which you expect to produce a good separation, based upon your past experience with chromatography. After you recover your fractions, take a sample right away. If you intend to save this protein for later use, dialyze it again. For all samples determine the concentration of the protein using the Bradford assay. Calculate the specific activity at every step, the total activity, the percent recovery, the purity factor, and the overall efficiency of the purification step.

PRE-LAB QUESTIONS

1. Prepare a flow chart for how you expect to purify L-lactate dehydrogenase, based upon your experience with salting out, and three different forms of chromatography. Select the best two types of chromatography to achieve the purification you want. What analytical tool will you need in order to perform the assay for your enzyme?

2. As in the previous question, prepare a flow chart for how you expect to purify catalase, based upon your experience with salting out, and three different forms of chromatography. Select the best two types of chromatography to achieve the purification you want. What analytical tool will you need in order to perform the assay for your enzyme?

POST-LAB QUESTIONS

1. You draw a protein out of dialysis. With a volume of 22.5 mL, a relative activity of 98.1 U/mL, and a concentration of 1.75 mg/mL, the protein has total units of activity equal to 2210 U and specific activity of 56.1 U/mg. You load this onto a chromatography column and collect the fractions which contain the most pure protein. In the pooled fractions you have 2030 U total activity and a specific activity of 275 U/mg.
   a. What is your % recovery from the chromatography step?

   b. What is your increase in purity?

   c. What is the value of this step as a purification step?

2. Often when you perform a purification step, you cannot account for all the units of activity that you started with. You might load 100,000 U onto a chromatography column, and only have 20,000 units in all the fractions that come off the column. Why do you lose activity with each step?

3. Theoretically, you should never see an increase in activity when you are performing a purification. However, sometimes a student will have 100,000 units of activity before they perform a certain step, and then have 150,000 units afterwards. What are the sources of error that would lead to apparent increases of activity? (You do not have to know anything about the enzyme or the purification step to be able to answer this question.)

10 Polyacrylamide Gels

ACRYLAMIDE CHEMISTRY AFFECTS SEPARATIONS

Gel filtration chromatography works on a principle that is connected to the topic of polyacrylamide gels, and also agarose gels, though they will not be discussed in this section. In gel filtration, there are beads of resin, and each bead has a number of cracks, pores, and crevasses. Any protein that is small enough to fit into these openings will get inside and remain stuck there for a certain amount of time, thus slowing its progress. The further it can get inside, the further stuck it will be, because it takes longer to get out again. Large molecules can just go around the beads, and don't get into the openings. Hence, they move through gel filtration columns faster.

Consider, though: the only reason that the larger molecules move faster through gel filtration columns is because they have the option of not going into the openings at all. What if that option were removed? What if there were not beads inside a column, but rather one single molecule of resin stretching throughout the entire space? This is a possibility when polymers are created inside the column, or any other three-dimensional mold. If the resin is not produced as beads, but rather as a single slab of material formed inside the column space, then there is no possibility of large molecules going around the resin. They must fit through the openings, or they will not move at all. Admittedly, this can be somewhat like trying to get a football through a garden hose, but with sufficient pressure behind it, even this feat is possible.

Acrylamide is a small three-carbon amide with a carbon–carbon double bond conjugated to the carbonyl. As a monomer it is carcinogenic, mutagenic, and a cumulative neurotoxin. Its structure and the structure of bis-acrylamide are shown in Figure 10.1.

Their double-bond structures makes them ideal for radical polymerization reactions. If an initiator attacks the terminal carbon, it results in a radical that is stable enough to react with another molecule of acrylamide at the terminal position (see Figure 10.2). Thus does polyacrylamide begin to grow. This reaction will propagate indefinitely, until finally being terminated by reaction with molecular oxygen, another

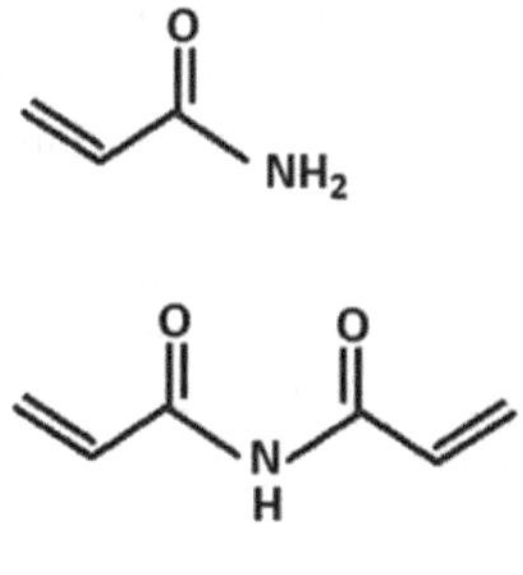

O
NH2

O O
N
H

FIGURE 10.1 Acrylamide and bisacrylamide monomers are the precursors of a polyacrylamide gel.

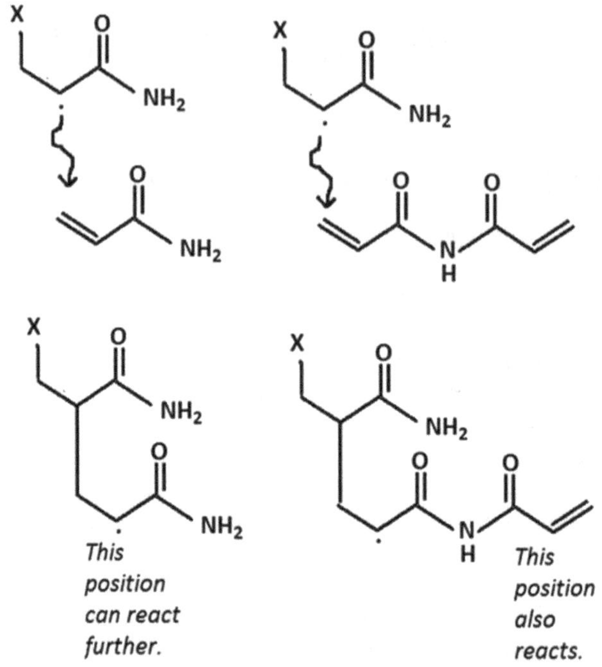

This position can react further.

This position also reacts.

FIGURE 10.2 The free radical at the alpha carbon position can attack the exposed beta carbon, creating a new free radical. This new free radical propagates the polymerization reaction. Bisacrylamide can propagate two separate polymerization reactions.

radical end of a polyacrylamide chain, or the wall of the container in which it is made. The resulting product will be long fibers of polyacrylamide stretching across or along the container, and it is called a kind of "gel". It has a "gelatinous" texture, you see.

Simple polyacrylamide fibers do not hold together well, and large molecules with a lot of pressure behind them will easily push the fibers apart. The large molecule would then quickly pass through the gel. It would not serve very well as a chromatography medium. However, there is a second chemical reaction that can be added in while the first polymerization reaction is occurring. Bis-acrylamide is a secondary amide made of two units of acrylamide joined at the nitrogen. Each side of the molecule is capable of independently participating in the acrylamide polymerization reaction. A growing polyacrylamide chain that adds a molecule of bis-acrylamide will become crosslinked to another polyacrylamide chain. By blending regular acrylamide and bis-acrylamide in certain proportions, you can control how often these crosslink events occur. For example, if you have a mole ratio of 1 bis-acrylamide to 14 acrylamides, then on average there will be one crosslink that occurs every 15 times (*i.e.* $1+14=15$). Similarly, if you have a ratio of 1 bis-acrylamide to 19 acrylamide, there will be a crosslink every 20 times.

These crosslink events determine the size of the holes in the gel. The gel which has a crosslink every 15 times has holes that are smaller than the one with a crosslink every 20 times. It will slow down large proteins more. Proteins much smaller than the holes would move very quickly and not be well separated. Proteins much larger than the holes would hardly move through them at all, and also not be well separated. It is the proteins whose sizes are somewhat on the scale of the holes created that will find a path through the gel, and be separated depending upon their sizes.

The other feature that will change the separation is the actual number of monomer fibers creating blocking features. This is the ratio of acrylamide to water, or the percentage of acrylamide in a gel. A wall of polymer with even very small holes will not slow down a large molecule if there is a path around the wall. A gel which is only 5% acrylamide has many more openings in it which very big molecules can find their way through, thus circumventing any holes. It will allow larger proteins to travel quickly, and small proteins to move without any restraint whatsoever. Thus, the smaller proteins would not be well separated. A gel that was 15% acrylamide might do a very good job of slowing these smaller proteins down, but it would be so dense that larger proteins would not be able to penetrate at all. In this case, there would be good separation of the small proteins, but poor separation of the larger ones. The percentage of total acrylamide affects separation just as the ratio of acrylamide to bis-acrylamide does, but the two things affect separation for independent reasons.

Thus, when casting a gel, you want to determine first the size of what you are separating, and cast a gel proportionately. You will modify two things: (1) the proportion of bis-acrylamide to acrylamide and (2) the percentage of total acrylamide in the gel. You cast the gel you want to achieve the best separation at the size range you desire.

BUFFERING GELS

Unlike column chromatography, you cannot achieve separation of macromolecules simply by flowing a buffer past the polyacrylamide gel. In this case, the buffer is not the mobile phase. The gel is too dense a material for a solution to flow through with any kind of current, and diffusion itself is faster than the solvent's current, anyways. That is not a good method for separating materials from a stationary phase and a mobile phase. In this case, the mobile phase is a current of electricity: if something the size of a protein is going to go through the little openings in the gel polymer, it is going to have to have some very strong pushing power behind it. Electrical potential offers the strength that is needed. The gel is put in what is essentially a large capacitor. The cathode (negative charge side) is on one end, and the anode (positive charge side) is on the other end. When an electrically conductive material is in contact with both the cathode and the anode, the circuit is complete and current begins to flow. The charge repulsion between cathode and the protein and the charge attraction towards the anode begins to push and pull the protein through the gel matrix with great force. It needs merely to have a good electrolyte solution in the space between the ends of the capacitor to create this effect.

That said, the protein and the gel need to remain intact while this electrical current is passing through. That requires the pH and the ion content to be maintained at some constant level. This is especially true when the scientist running the gel hopes to recover the protein intact from the gel.

You will want to have a lower pH buffer in the stacking portion of your gel and a higher pH buffer in the resolving portion of the gel. The difference in charge that it induces causes the proteins to form a tight band before beginning to resolve, and thereby improves the quality of the separation.

Goal

(Week 1) In this lab, you will prepare a 10% SDS-polyacrylamide gel, and store it for later use.

Method

1. First prepare your ingredients.
   Tris-Glycine Buffer, pH 6.5
   Tris-Acetate-EDTA Buffer, pH 8.5
   Deionized water
   19:0 acrylamide/bis-acrylamide solution
   TEMED
   10% ammonium persulfate
   Clean gel casting apparatus, including plates, clamps, bottom gasket, side spacers, and gel combs.
   Glass transfer pipets and rubber bulbs.
   15 mL and 50 mL disposable conical tubes
   1-butanol
   Protective gloves
2. Do not mix the ingredients but determine the volumes of each component that you will need. Determine if your protein's mass is more appropriate for the lower or the higher concentration stacking and separating gels.

Reagent	Volumes for 5% Stacking Gel	Volumes for 7.5% Stacking Gel	Volumes for 10% Separating Gel	Volumes for 15% Separating Gel
4× separating buffer	None	None	6.25 mL	6.25 mL
4× stacking buffer	1.25 mL	1.25 mL	None	None
Deionized H_2O	3.1 mL	2.775 mL	12.5 mL	9.375 mL
Acrylamide/Bis-acrylamide	0.65 mL	0.975 mL	6.25 mL	9.375 mL
10% ammonium persulfate	50 μL	50 μL	82.5 μL	82.5 μL
TEMED	7.5 μL	7.5 μL	15 μL	15 μL
Total Volume:	5 mL	5 mL	25 mL	25 mL

3. Clean the glass plates very thoroughly, using soap and water. Dry them thoroughly. Set up the gel caster apparatus so that your glass plates are pressed firmly against the rubber bottom gasket, with the side spacers clamped into place. Pour water into the casting apparatus to make sure it is not leaking. Then pour the water out the top, and somewhat dry the casting apparatus using Kimwipe tissues. If the apparatus leaks, you must reassemble it so that it does not leak before you begin. Place the comb in, and use a sharpie marker to mark out a line approximately half a centimeter below the bottom edge of the comb teeth. Then remove the comb and set it to the side.

4. Prepare 1 mL of 10% ammonium persulfate, and keep it on ice. This substance is actually unstable in water, and cannot be used after an hour, even if it is kept on ice. This time will be greatly shortened if it is not kept on ice.
5. Blend the materials for the separating gel in a 50 mL tube, in the following order: buffer, water, acrylamide, ammonium persulfate. Then pause and gently mix these components, drawing them in and out of a transfer pipet.
6. Next add the TEMED and quickly but gently mix with the transfer pipet. This initializes the polymerization reaction. Use the transfer pipet to add the solution to the inside of the casting apparatus. Add this solution up to the line you marked. Using a different transfer pipet, overlay the acrylamide solution with a later of 1-butanol. Dispose of the transfer pipet, and let any remaining acrylamide in the 50 mL tube polymerize. You will need to wait for about 15 minutes for the reaction to proceed before going on to the next step.
7. Gently tip the casting apparatus sideways to determine if the polymerization reaction proceeded. You should see the butanol move and a solid gel not move. Pour off the butanol, and rinse with deionized water, which you also discard.
8. Blend the materials for the stacking gel in a 15 mL tube, in the same order you did for the separating gel. Mix the components as before.
9. Add the TEMED to initialize this second polymerization reaction. Use the transfer pipet to add the solution on top of the separating gel you created. Add to about a half centimeter from the top of the casting plate. Then, insert the gel comb. Dispose of the transfer pipet and let the remaining acrylamide polymerize inside the 15 mL tube, just as you did before. You will need to wait 15 minutes, as before for the reaction to polymerize.
10. Gently detach the comb, and remove it. You should see clear edges for wells. (Note: if there are many combs available, then it is better practice to partially remove the comb, just enough to check that the wells are present, but then to reinsert the comb, and store the gel with the comb in it until you are ready to use the gel.)
11. Remove the casting plates from the casting apparatus. Do not remove the spacers or the glass plates: the gel should be in between the casting plates. Wrap the entire gel in wet paper towels, and place inside an airtight plastic bag. Store at 4°C until you are ready to use the gel. You should use it within a couple of weeks or it will not be fresh.

LOADING AND RUNNING AN SDS-PAGE GEL*

GOAL (WEEK 2)

In this lab, you will load the wells of a polyacrylamide gel, resolve the components, and image them. Then you will determine the size of different protein samples, using the relative mobilities of standards as guides.

* A subsequent lab, or a standalone.

Method

1. First prepare your proteins. You will need protein standards that range between about 70 kDa and 12 kDa. Also prepare the following proteins at a concentration of 1 to 2 mg/mL: casein, albumin, cytochrome C, hemoglobin, riboflavin-binding protein, and trypsin. Other proteins with a monomer mass between 12 and 70 kDa may be substituted for any of these.
2. Prepare your other reagents:
   25 mL of 50:50 glycerol/water mix
   Tris-Acetate-EDTA buffer with 10% SDS, pH 8.5
   A vertical gel apparatus
   Power source
   2x loading buffer with β-mercaptonol and 1% bromophenol blue
3. In a vertical gel apparatus, you must build an electrophoresis chamber. In a holder, you must place a gel facing inwards such that it makes a firm seal to the rubber gasket of your holder. If there is any electrical resistance tape on the gel, you must remove the tape prior to putting it into the holder. If there is a comb to help shape the wells, this must be removed as well. (Both are common with commercially available gels.)
4. You must place either a second gel or a buffer dam on the other side of the holder, again making sure that there is a firm seal to the rubber gasket.
5. Place the holder into an electrophoresis reservoir. Pour 1x Tris/SDS buffer into the center of the holder and into the outside reservoir. It is best to let the buffer run over the edges of the holder and into the reservoir below. The reservoir at the center of the holder must be completely filled, such that the level of the buffer is *above* the top of the gel. The outside reservoir must be filled sufficiently to allow contact with the cathode. For a typical minigel, this will take between 600 mL and 900 mL of buffer.
6. Wait and observe the apparatus. If the buffer in the inner reservoir is leaking, the apparatus must be disassembled and remade with firm contact to the rubber gasket. Return to Step 1, if so. Otherwise, proceed on to Step 5.
7. Wash the wells flushing them with buffer. This can be done by pipetting buffer from the inner reservoir in and out of the wells, using a glass pipet or a 1000 mL pipettor.
8. Prepare your samples. For a minigel, this is done by mixing in an Eppendorf tube 6 μL of protein with 6 μL of 2X loading dye. It is also recommended that you add to this volume 5 μL of 50:50 glycerol/water mix. Though this skews the buffer concentration to slightly less than 1X, it makes the sample settle better in the next step. If you do know your protein concentration (as in Step 1), the optimal amount to add per well is 10 to 20 μg, so you should try to have between 17 and 34 μg of protein in the 6 μL of protein you add. If you do not know your protein concentration, and have extra wells available, then it is a wise idea to make serial dilutions. After you mix the components, centrifuge the Eppendorf tubes for 20 seconds at 8,000 × g to 10,000 × g to settle all the liquid at the bottom of the tube.

9. Starting in the well at the left, you should put whatever molecular weight standards you are using. These are often proteins of known molar mass, whose values lie both above and below the protein you are studying. If you do not know the mass of your protein, it is advisable to use a broad range of markers, possibly using both the first two wells. Often, if you have a purified specimen of your protein of interest, you use a well with this protein as a standard. In the standards' well(s) you should pipet 10 μL using a very fine needle pipet.
10. Into each well from the left to the right, pipet 10 μL of each sample into each well. If you added the glycerol, it invariably sinks quickly and you can see it form a layer inside the well. If you did not add glycerol, you often must add it slowly to make sure that it stays in the well.
11. When you have loaded your wells, attach the electrodes of the electrophoresis chamber to the power supply. Be certain to attach the cathode end of the electrode to the cathode end of the power supply and the anode to the anode. Otherwise, your current will run backwards and the proteins will not be impelled into the gel. (In some nationalities, this is color coded so that red wires attach to red outlets and black wires attach to black outlets.)
12. Initialize the power on the electrophoresis unit and run it at constant current of 130 to 150mA. (Alternatively, you can run the gel with constant voltage of 180 V to 200 V.)
13. When the gel has finished running after approximately 30 to 40 minutes, turn off the power to the electrophoresis unit.
14. Remove the gel from its casting frame and prepare it for imaging. If you are using a commercial Stain-free gel, you need merely place it on a Stain-free imaging plate and illuminate it with the right frequency of light. Image the gel using the appropriate light source and CCD camera. If you are blue staining or silver staining, then you must soak the gel in the appropriate stain for 15 to 30 minutes, remove the gel from the stain solution, wash off the excess dye using deionized water, then add destain solution and destain overnight, changing the destain solution at least once. For Coomassie Blue stain, the addition of a few paper tissues (especially Kimwipe tissues) will speed the process immensely.
15. Measure the distance the standards have moved and develop a standard curve by the relationship

$$\ln(\mathrm{MW}) = \mathrm{m} * \mathrm{R_f} + \mathrm{b}$$

where MW is the molar mass, R_f is the relative mobility (distance spot travelled/distance dye front travelled), *m* is the slope, and *b* the y-intercept.
16. Determine the molar mass of your bands by comparison to the standard curve. Bands that are significantly outside the range of the standard curve will not be correctly measured. Likewise, if the standards are too far apart in molar mass, they will not make a linear standard curve (see Figure 10.3).

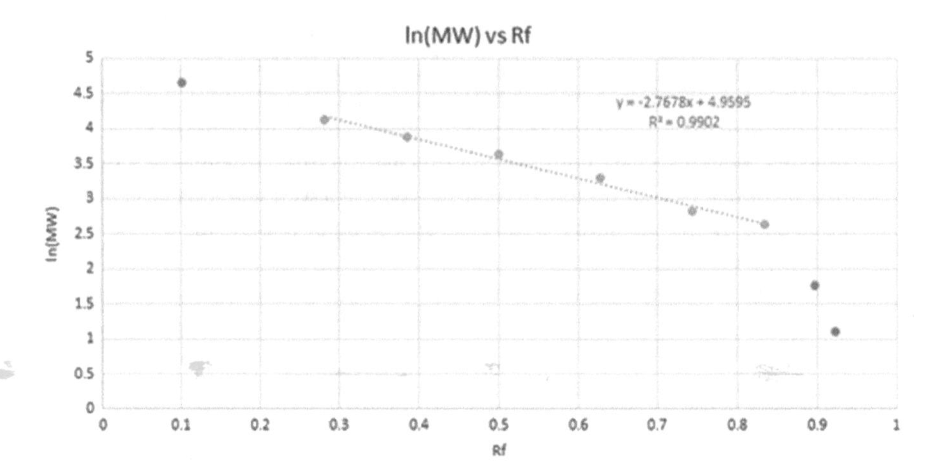

FIGURE 10.3 Typically, when ln(MW) is plotted against Rf for the standards, only some of the data will fit a linear plot. The molar masses of protein samples within this range can be accurately determined, but the error for proteins outside this range increases rapidly.

PRE-LAB QUESTIONS

1. You are running an SDS PAGE gel. You load the second column with some standards, then the next four with cytochrome C (UniProt # P99999), Hemoglobin (#P69905 for alpha and P68871 for beta), Lysozyme (#P61626), and a 1:250 Trypsin blend (#P07477), but not necessarily in that order.

   Your standards have molar masses (in kDa) of 3.0, 5.9, 14, 17, 27, 38, 49, 62, and 105. You run a gel, and it looks like the one in Figure 10.4.

   The lines at the top of the gel are some graphite powder that your professor recommended you add to your samples so that the starting point in the

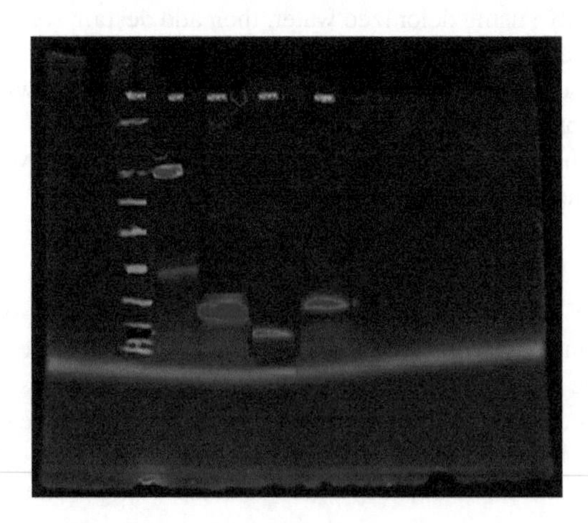

FIGURE 10.4 Use this image to determine the relative mobilities of the standards and the samples.

wells would be imaged more clearly. Graphite is not water soluble, and can be added to the wells after the gel has been run; it settles into the well and makes the starting point clearer.

In order to analyze this gel, do the following:

a. Identify which line corresponds to which standard. Measure the distance the solvent front moved in the standards' lane (d_s). Measure the distance each of the standard proteins moved (d_x). Calculate R_f for each standard.

$$R_f = d_x / d_s$$

Then calculate ln(MW) for each standard

MW (kDa)	d_s	d_x	R_f	ln(MW)
105				
62				
49				
38				
27				
17				
14				
5.9				
3.0				

b. Using some graphing program or by hand, if that is your strong preference, plot ln(MW) versus R_f. Identify which data points are linear and which curve away from the line. There will *always* be data points in your standards that deviate away from the ones that are linear, either at the highest molar masses, or at the lowest, or at both the highest and the lowest. For the ***linear points only***, determine the equation which describes the relationship between ln(MW) and $R_{f+.}$

c. For columns A, B, C, and D, identify the major bands that appear in each column. Ignore mere debris or shadows in a column. There are never more than two bands in any of these columns, and only one in most cases. Refer to them as “A1”, “A2”, “B1”, “C1”, and so forth. Measure d_s and d_x for each of the major bands. Calculate R_f for each band. Then calculate ln(MW) using your equation above. Finally calculate MW of the protein in each band.

Name	d_s	d_x	R_f	ln(MW)	MW (kDa)
A1					
A2					
B1					
C1					
D1					

d. Determine the Identity of the bands that correspond to Cytochrome C, Hemoglobin, Lysozyme, and Trypsin by comparison to the molar masses given in the UniProt database.

Name	Identity	MW (kDa)	UniProt MW
	Cytochrome C		
	Hemoglobin		
	Lysozyme		
	Trypsin		

e. The trypsin is blended with an innocuous protein. Determine whether this contaminant is Serum Albumin.

POST LAB QUESTIONS

1. Why is it necessary to use only the standards which show a direct proportion between R_f and ln(MW) when generating your standard curve? Can you accurately determine the molar mass of any protein outside this range on your gel?

2. If you decrease the amount of acrylamide in your gel, which proteins will move more quickly: higher molar mass proteins or lower molar mass proteins? Should it shift the "linear" region of your standards to the lower molar mass range, or the higher molar mass range?

3. If you use a smaller ratio of acrylamide to bis-acrylamide when casting your gel, will it increase or decrease the separation between high molar mass proteins and lower molar mass proteins. Explain your answer.

11 *In Silico* Biochemistry *The Evolution of Globins*

There is a large branch of modern biochemistry that is entirely computational. So-called "dry biochemistry" is as important a tool in the biochemist's kit as any "wet lab" technique.* To utterly master the skills of a computational biochemist require intensive study of mathematics and computational science as well as regular biochemistry theory. Consequently, most computational biochemists are not wet lab biochemists, but rather work hand in hand with them, and vice versa. The best research teams always involve a mixture of computational and non-computational scientists.

Because the field of computational biochemistry requires years of mastery, this lab is only intended to give an idea of the most basic kind of work that can be done. True masters can map the sequence of a protein with unknown structure onto the known structure of a related protein, and predict the structure of the unknown one. They can determine the most probable transition state analogs, or determine probable binding sites. They can predict the probability of one protein associating with another, sometimes even the affinity of binding. However, in this lab, you will only be asked to determine the evolutionary distance of some gene separation events based upon one known separation event and the differences in primary structure of proteins from different genes.

GLOBINS

Globins are fairly small proteins that are very rich in alpha helical structures. The first protein structures ever solved were myoglobin and hemoglobin, both members of this family. Globin genes are quite common, and serve a large range of functions, but structurally tend to all be similar. The truth is that they are similar because they all derived from some common ancestor globin gene, long ago in time, making them *homologs*. Some of the most famous globins include myoglobin, and hemoglobin, as mentioned, but also neuroglobin and others.

The protein hemoglobin, shown in Figure 11.1, is actually a tetramer composed of two dimers, and each monomer of the dimer is from a different globin protein, related to each other but not the same. One of the monomers is referred to as alpha hemoglobin, and the other is beta hemoglobin. Neither one has any beta sheet structure: the names are just used to designate two different types of monomers

* The terms "wet" and "dry" commonly refer to work done at a lab bench or not at a lab bench, respectively. They are commonly used jargon terms in most labs. "*In vitro*" or "in vivo" also pertain to lab bench work, whereas the term "*in silico*" refers to computational work.

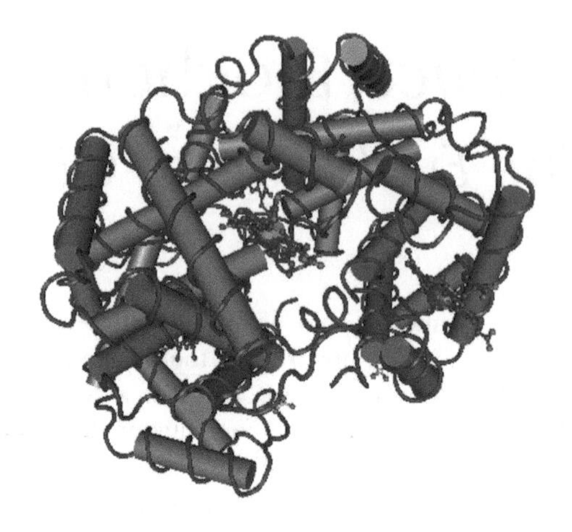

FIGURE 11.1 Human hemoglobin.

and come from an age when it was still not clear how many different globins there were going to be. Other types of hemoglobin monomers include gamma hemoglobin and "f" hemoglobin, which shows up in fetal blood cells. Each comes from a different gene, and each has a different sequence of amino acids for its primary structure.

BLOSUM

The computational tool BLAST, or "Basic Local Alignment Search Tool", is a way of calculating whether two sequences have similarity that exceeds what would be expected by random chance. If a large number of closely related sequences are aligned such that one residue matches up to another in a different sequence, there will be a large number of residues at the same position that will have the same identity. At the very least, they will have a residue with very closely related identity, such as leucine or isoleucine, or alanine and valine. For example, in the proteins lactate dehydrogenase B and lactate dehydrogenase C from *bos mutus* (the water buffalo) the identity of the residues from positions 331 to 336 is "IQKDLK" in the former, whereas positions 339 to 344 have the identity "VQKDLK"" in the latter. Notice that they are identical, except for the first residue in the sequence, isoleucine in LDH B and valine in LDH C. However, isoleucine and valine are extremely similar compounds in all their physical properties, and differ sterically only by the presence of an extra methylene group. This is a good match. If one examines the residues on either side of these ones in both proteins, it soon becomes apparent that LDH C is virtually the same sequence, shifted by eight residues, and eight residues have been added in earlier in the sequence. This is therefore a very close match. All other sequences that align with closely matched residues are considered related, because they defy the probability of getting that same sequence in the same positions by mere random chance. Large families of sequences are assembled this way.

In 1992, the Henikoffs developed a computational tool for evaluating the probability that any individual amino acid residue in a primary structure was going to be replaced by some other amino acid residue. This tool is referred to as BLOSUM, which is an acronym from "Block Substitution Matrix". Their first step was to assemble many large arrays of related sequences and determine the statistical probability that any one residue would be replaced by some other residue. In point of fact, they used over 2000 blocks of aligned sequences and over 500 families of proteins. Their probability system works as follows:

Imagine a simple three-residue sequence. The positions can be considered as M, W, X, Y, or Z. Now an aligned family of this three-residue sequence is found, and the variations observed are:

```
XYZ
ZYM
WYM
WYZ
XYW
```

We consider each column individually to determine the probability that one residue can be replaced by another. The first column is

```
X
Z
W
W
X
```

Similarly, we can examine the second column

```
Y
Y
Y
Y
Y
```

And so on with the third, until we have the alignment of each column. Then we compare what possible pairs we have. If X were found to be X in another sequence, it would be an "XX" pair. If it were found to be Z, then it would be an "XZ" pair. In fact, since it is arbitrary that X should go first, the "XZ" pair and the "ZX" pair are considered to be the same replacement, and are grouped together. Either one indicates that an X and a Z have been swapped out. A similar logic is made for every other pair. The pairs of each type are counted for every column and the frequencies placed into a table, where their total number is determined from all columns

Pair Type	Freq. Col 1	Freq. Col 2	Freq. Col 3	Total Number
XX	1	0	0	1
XZ	2	0	0	2
XW	4	0	0	4
YY	0	10	0	10
ZZ	0	0	1	1
ZM	0	0	4	4
ZW	2	0	2	4
MW	0	0	2	2

Notice that some combinations do not appear at all. There is no "XY" pair anywhere in this set, and no "ZY" pair, for example. We omit them from the calculation, given that they have a frequency in this set of zero. Notice also that YY appears with a frequency of ten, because the first Y has four cases where it is also Y, the second has three, the third has two, and the fourth has one, for a total of 10 instances.

We are now capable of making a matrix that gives the probability of any letter being substituted for another one:

	M	W	X	Y	Z
M	0				
W	2	0			
X	0	4	1		
Y	0	0	0	10	
Z	0	4	2	0	1

The values in this matrix come from the totals in the frequency chart. If the value is zero, that pair never showed up at all in the frequency chart. We then use a statistical analysis to determine the probability of any letter being replaced by another letter in the sequence. A description of the analysis is appended to this section, but knowing how to get it is different than knowing how to use it. What we can see, even without the numerical probability scores, is that "Y" is not replaced by any other letter. Somehow, it is quite resistant to being replaced, and tends to retain its identity. On the other hand, "X" and "W" are more likely to be replaced by each other than they are to retain their identities. In a longer sequence of these letters, we could expect to find them replacing each other often.

The same analysis was done on a much larger scale by the Henikoffs for proteins, and it is to be hoped that the scale of the work of 2000 alignments of 500 residues with 20 possible amino acid residues at any given position is appreciated by all users of the BLOSUM matrix. The table they generated is known as the BLOSUM62 substitution matrix, based on 62% identity of aligned structures for those used to calculate the matrix. Other BLOSUM matrices have also been calculated, but increasing the identity of source alignments does not seem to improve the probability matrix any further than 62. The scores of a given pair exchange are shown below. In this matrix, the probabilities have been normalized, so that a score of "0" indicates the median preference of a substitution occurring, positive value indicating more preference for the particular substitution than the median and negative values indicating less preference. In other words, the more positive the value, the more frequently that pair is observed. The more negative the value, the less frequently.

BLOSUM 62 Substitution Matrix

	C	S	T	P	A	G	N	D	E	Q	H	R	K	M	I	L	V	F	Y	W
C	**9**	−1	−1	−3	0	−3	−3	−3	−4	−3	−3	−3	−3	−1	−1	−1	−1	−2	−2	−2
S	−1	**4**	1	−1	1	0	1	0	0	0	−1	−1	0	−1	−2	−2	−2	−2	−2	−3
T	−1	1	**4**	1	−1	1	0	1	0	0	0	−1	0	−1	−2	−2	−2	−2	−2	−3
P	−3	−1	1	**7**	−1	−2	−2	−1	−1	−1	−2	−2	−1	−2	−3	−3	−2	−4	−3	−4
A	0	1	−1	−1	**4**	0	−2	−2	−1	−1	−2	−1	−1	−1	−1	−1	0	−2	−2	−3
G	−3	0	1	−2	0	**6**	0	−1	−2	−2	−2	−2	−2	−3	−4	−4	−3	−3	−3	−2
N	−3	1	0	−2	−2	0	**6**	1	0	0	1	0	0	−2	−3	−3	−3	−3	−2	−4
D	−3	0	1	−1	−2	−1	1	**6**	2	0	1	−2	−1	−3	−3	−4	−3	−3	−3	−4
E	−4	0	0	−1	−1	−2	0	2	**5**	2	0	0	1	−2	−3	−3	−2	−3	−2	−3
Q	−3	0	0	−1	−1	−2	0	0	2	**5**	0	1	1	0	−3	−2	−2	−3	−1	−2
H	−3	−1	0	−2	−2	−2	1	1	0	0	**8**	0	−1	−2	−3	−3	−3	−1	2	−2
R	−3	−1	−1	−2	−1	−2	0	−2	0	1	0	**5**	2	−1	−3	−2	−3	−3	−2	−3
K	−3	0	0	−1	−1	−2	0	−1	1	1	−1	2	**5**	−1	−3	−2	−2	−3	−2	−3
M	−1	−1	−1	−2	−1	−3	−2	−3	−2	0	−2	−1	−1	**5**	1	2	1	0	−1	−1
I	−1	−2	−2	−3	−1	−4	−3	−3	−3	−3	−3	−3	−3	1	**4**	2	3	0	−1	−3
L	−1	−2	−2	−3	−1	−4	−3	−4	−3	−2	−3	−2	−2	2	2	**4**	1	0	−1	−2
V	−1	−2	−2	−2	0	−3	−3	−3	−2	−2	−3	−3	−2	1	3	1	**4**	−1	−1	−3
F	−2	−2	−2	−4	−2	−3	−3	−3	−3	−3	−1	−3	−3	0	0	0	−1	**6**	3	1
Y	−2	−2	−2	−3	−2	−3	−2	−3	−2	−1	2	−2	−2	−1	−1	−1	−1	3	**7**	2
W	−2	−3	−3	−4	−3	−2	−4	−4	−3	−2	−2	−3	−3	−1	−3	−2	−3	1	2	**11**

This table presents some surprises. One would expect structural or charge similarity to allow the most frequent substitutions, but this is not always what is seen. Most biochemists group the amino acids as follows:

G,A,V,L,I, M aliphatic (though some would not include G)
S,T,C hydroxyl, sulfhydryl, polar
N,Q amide side chains
F,W,Y aromatic
H,K,R basic
D,E acidic

Considered this way, there are some rather odd substitution frequencies. For example, there is equal preference for asparagine to histidine pairs and aspartate pairs, even though asparagine and aspartate are sterically quite similar, and histidine differs both in charge group and steric occupancy. This suggests that what evolution thinks is "similar" is not necessarily similar to the molecular biologist or biochemist. Clearly, both substitutions result in equally functional products in many cases.

ALIGNING AND SCORING SEQUENCES

Any two sequences of proteins can be aligned, whether they are certainly related or not. A sequence can even be scored against itself, and usually is, as a reference value. The researcher wishes to align them in such a way as to give the best possible score value, described as follows, and more closely related sequences will score more highly.

Scoring is fairly simple: every single pair is given the score from the BLOSUM table. Then the total score of the pairs is added. Consider the score of the following short peptide against itself, to create a reference value:

```
MRLLV
MRLLV
```

The pairs receive the scores, in order, of 5, 5, 4, 4, and 4, for a total of 22. This is the highest score that any other sequence could ever achieve when being scored against this particular peptide. Now consider how it scores against a related peptide, MKLLL.

```
MRLLV
MKLLL
```

The scores of the pairs are now 5, 2, 4, 4, and 1, for a total of 16.

GAP PENALTIES

As stated, the peptides should be aligned in such a way that the total score is the highest possible. Sometimes this requires the researcher to accept that gaps appear in sequences, either because some amino acids have been added, or because a few have been deleted. Biologically this has happened without the loss of function of a vital protein, so addition or deletion events are not always negative occurrences. They do, however, indicate a certain amount of evolutionary distance, and so cannot be considered as "null events" which get scores of "0" in every case. There is some penalty that is factored into the scoring. There are a number of methods for calculating the gap penalty

1. Linear: In this case, every gap is considered a negative event, and is assigned a value of "–1". Certainly the easiest penalty to calculate, it also is the worst approximation of reality. Consider the sequences:
   ```
   MRLL----V
   MRLLAATTV
   ```
   Notice that this is the same sequence as the reference we used before, but with two added alanines and two added threonines in-between two of our residues. The matches would score as before for a total of 22, but the mismatches would penalize the score by four so that the total is only 18. The linear penalty assumes that longer gaps are less closely related than shorter gaps.

2. Linear: In this case, the number of gaps is considered more negative than the length of the gaps. It is as easy to calculate, and slightly better at approximating evolutionary distance: an addition or deletion of a long loop of amino acids could occur in a single event, but five different additions or deletions would have to have occurred in five different events. The same sequences can be compared:
   ```
   MRLL----V
   MRLLTTAAV
   ```
   This time the five matches give 22 points, and the single gap gives a −1 penalty, for a total score of 21. This indicates a closer relationship than was calculated by the linear method.
3. Affine: This is the most widely used penalty determining method. It calculates a penalty for opening a gap at all (p) and a penalty for the length of the gap (q) by n residues. The penalty is calculated as "$p+(n-1)q$". The difficulty comes in assigning values for p and q. The longer a sequence is, the likelier it will be that gaps will exist due to entirely random reasons, which gives p a lower value. Likewise, the size of the protein affects how long one would expect a random gap to be, thus lowering q for larger proteins. Lastly, there is a reciprocal relationship between the existence of a non-random gap and how long the gap is, thus lowering p if q is large and vice versa. This makes the affine method the hardest possible of the three to calculate, but it certainly gives the most accurate predictions of evolutionary distance. For purposes of this lab, it will give tolerably good results if p is always given a value of "−2" and q is given the value of "−0.2", though these values will certainly not work when considering other proteins than the ones assigned here.

Other methods exist as well, but will not be treated in this text.

ALIGNMENT TOOLS

There are a number of tools available to you to help you achieve the optimal alignment. The description above was to help you know how the tools work. One of the best tools is the BLAST tool offered from the National Center for Biological Technology at the National Institutes of Health (www.ncbi.nlm.nih.gov). All the tools available for analysis can be found on a subpage (www.ncbi.nlm.nih.gov/home/analyze/), and you should select "BLAST Link (Blink)". Next, out of the options you are given, choose "Protein Blast". You will be asked for the accession number or FASTA sequence. The accession number or FASTA sequence can be obtained via the Protein Webpage at the NCBI (www.ncbi.nlm.nih.gov/protein/), where you can search for your protein by name and taxonomy. The accession number is displayed with the search results. To compare two sequences, enter the accession number or FASTA sequence of the reference protein in the space provided, then select the button "Align two or more sequences". It will create another box for providing a new accession number/FASTA sequence. Enter this term for the second protein. Then click on the "BLAST" button. When the calculations are complete, you should see displayed any sequences that align and the optimal position of any gaps that are present. If there is absolutely no alignment better than random chance, then it is not shown: the two sequences can just be lined up next to each other with no attempt at finding the best position, because no better score can be found than this.

CALCULATING EVOLUTIONARY DISTANCE

Calculating the scores of two protein sequences may be quite complicated and difficult to do, but having done that, calculating the evolutionary distance is not difficult at all. It reduces to a simple formula:

$$\text{Evolutionary Distance (unitless)} = 1 - \text{score}_{comp}/\text{score}_{ref}$$

where score_{comp} is the score for two different sequences compared to each other and score_{ref} is the score of your reference sequence to itself. For our examples used above,

```
Reference sequence: MRLLV
Test sequence: MKLLL
```

our score_{ref} was 22 and our score_{comp} was 16. The evolutionary distance between these is calculated as

$$\text{Evolutionary Distance (unitless)} = 1 - (16/22) = 0.27$$

The smaller the value, the more closely related two sequences will be. If one determines the distance between several sequences, it is possible to find out how long ago they diverged from the one that is being used as the reference. For instance, if sequences B, C, D, and E have evolutionary distances from A with values of 0.80, 0.55, 0.40, and 0.20, they can be diagrammed as shown in Figure 11.2.

This is the first step in creating a cladogram, a visual representation of the evolutionary distance between genes or organisms. In this case, it is revealing the distance between genes, because the protein sequences show genetic separation, not species separation. Notice that the branches occur at distances which are additive: the branch at which D separates from the A&E lineage is 0.20 units before the separation between A and E, for a total distance of 0.4 units; then it is a further 0.15 units until the lineage that separates C, for a total distance of 0.55.

To refine the cladogram, the evolutionary distances are calculated a second time, this time using sequence E as the reference. In this case, the distances of E to A, B, C, and D, are found to be 0.20, 0.80, 0.60, and 0.35. This creates a second set of distances, shown in Figure 11.3, but there are some differences as well.

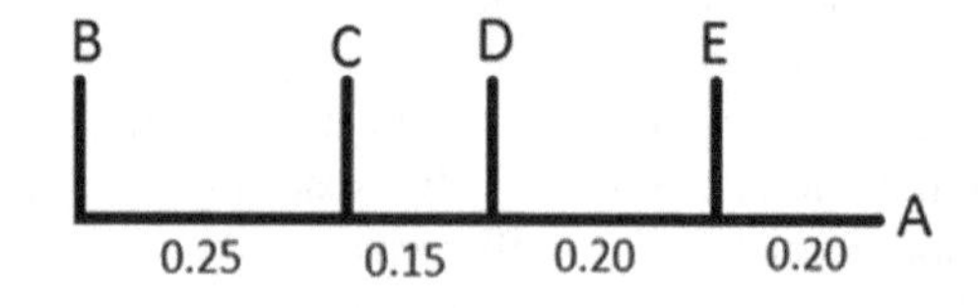

FIGURE 11.2 An initial cladogram between three sequences.

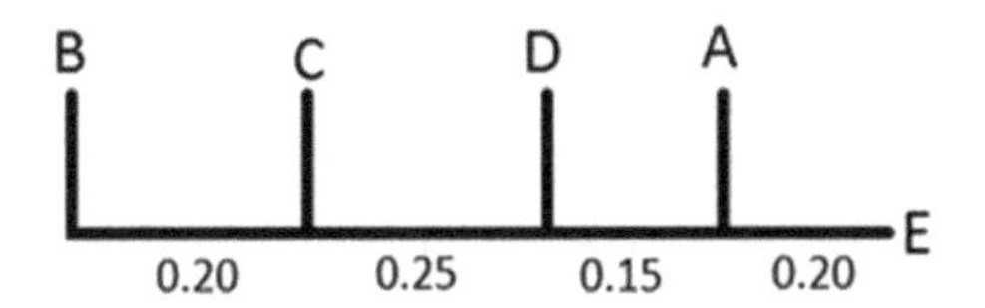

FIGURE 11.3 A different cladogram between sequences is formed when a different sequence is considered the reference. A number of genetic events can lead to this occurence.

We notice that the gap between E and C is slightly larger than the gap between A and C, for instance, indicating that A is slightly more closely related to C. Similarly, the gap between E and D is slightly smaller than the gap A to D, indicating that E is the closer relative to D. Discrepancies like this give a means to calculate where branches of a family tree are with greater probability. The length of the branches can be calculated as follows.

There is some separation between the lines that lead to A, E, and D. The distance between the origin of separation and A can be designated with the variable x, the distance between the separation and E can be designated with y, and the distance between the separation and the separation event to D can be designated with z, as shown in Figure 11.4.

We know some pieces of information:

$$x + y = 0.20 \text{(from both cladograms)}$$

$$x + z = 0.40 \text{(from cladogram 1)}$$

$$y + z = 0.35 \text{(from cladogram 2)}$$

Mathematically, this allows us to give both x and y as functions of z:

$$x = 0.40 - z$$

$$y = 0.35 - z$$

These two terms can be substituted in the first relationship:

$$x + y = (0.40 - z) + (0.35 - z) = 0.75 - 2z = 0.20$$

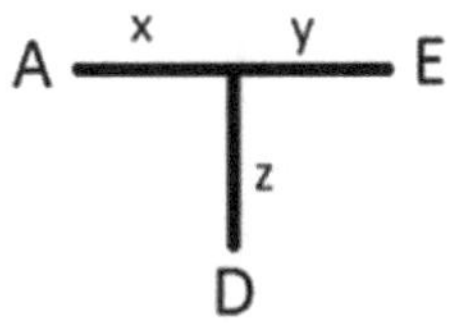

FIGURE 11.4 With sufficient measurements, it is possible to determine how far the separation is between sets of sequences.

Algebraically,

$$2z = 0.75 - 0.20 = 0.55$$
$$z = 0.55/2 = 0.275$$

With this term, we can solve the distances x and y:

$$x = 0.40 - 0.275 = 0.125$$
$$y = 0.35 - 0.275 = 0.075$$

This gives us our new cladogram, displaying the unequal separation of E and A since they separated from the lineage of D.

We can repeat this process with A, D, and C to find the distance of the new separation, shown in Figure 11.5. It is better to use gene separations that occurred on more comparable timescales. Thus, it makes more sense to find the separation between A, D, and C than between A, E, and C, because A and E are so close to each other that their lineage might appear to be only one event compared to the separation between the lineage of C versus A. Since D is more widely separated from A, it is easier from the point of view of the C lineage separation to score it as a different event than the separation of A. Thus, it would be more reliable to find the divergence between A, D, and C if we scored first the distances of all the genes from D, and compared the distance between A and C in the first cladogram to the distance from D and C in the third cladogram. This is similar to what was done above in finding the distance between A, E, and D using the first two cladograms.

Sometimes two different references will give contradictory values. It might be the case, for example, that if sequence C were used as a reference, sequence A might seem closer. In this case, one looks at the actual value of evolutionary distance, and the distance that scores as the closer evolutionary relationship is generally more trustworthy. Such discrepancies often occur when the lineage of a gene also includes some lateral transfers. An example of this can be found in certain genes between wolves, domestic dogs, and coyotes. Although domestic dogs are closer relatives of wolves than coyotes are, certain genes from the coyote line show up closer to the wolves. This is due to lateral interbreeding between the species in some rare cases. In other cases, especially in the yeast or microbial world, genes for proteins have been absorbed into the genome directly, without any interbreeding. They have created descendants that show closer relationships to certain forbear genes than would otherwise have been expected.

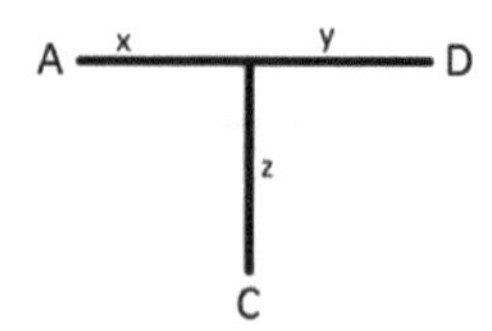

FIGURE 11.5 It is easier to find evolutionary separation with more divergent sequences than closely related ones. Sequence D is further removed from A, and so lends itself more to this type of analysis.

If one wants to determine the distance in time, as well, then there has to be some reference point where you know two protein sequences diverged from one another at a certain point in time. That reference then is used to calculate proportionally how many years passed. For example, suppose that the reference was scored against another protein "C" and had an evolutionary distance of 0.33. In this case, though, you have evidence that suggests that the reference protein and protein "C" drifted apart 330,000 years ago. Then you could use simple proportionality to find out how long ago a score of 0.27 occurred:

$$\text{Evolutionary Distance (years)} = 330{,}000 \text{ years} * (0.27/.0.33)$$

$$= 270{,}000 \text{ years.}$$

It must be noted that the scores can only come from proteins that are known to be related, if proportionality is to work. Additionally, there is a larger error when the time gap gets larger, indicating less certainty for very long evolutionary distances. This is because genes change in their rate of divergence over time if multiple copies of the gene are made or eliminated (*e.g.* number of amylase genes between humans and chimpanzees), or if other genes evolve to fulfill vital functions (*e.g.* different enzymes that perform the PEP carboxykinase activity with different substrates), or if a previously vital function becomes less important (*e.g.* L-gulonolactone oxidase activity in an organism that does not lack ascorbic acid in its diet), or if a previously unimportant function becomes vital (*e.g.* production of melanin or other pigments in a location whose atmosphere blocks less UV radiation). As time goes on, the probability of one or more of these events occurring increases. This will, in turn, change the rate at which genetic drift occurs.

Sources of error in calculating evolutionary distance will include (1) the length of the reference protein, which overstates the importance of any change the longer it is; (2) the identity of the reference protein, since any individual reference completely overlooks the possibility of lateral gene drifts, (3) the length of the target protein, which understates the importance of any change the longer it is; (4) the certainty of the reference time, which increases the uncertainty of every other calculated time.

Goal

In this experiment, you are to find the evolutionary distance between certain gene sequences in Table 11.1. You instructor may also assign a different set of sequences, if some have become recently relevant.

1. Score human hemoglobin beta against itself for a reference score.
2. Score all the other sequences against human hemoglobin beta.
3. Identify which sequences are so close as to show no evolutionary divergence. It would be expected that either or both of chimpanzees or bonobos might fall into this category.
4. Create a cladogram to indicate the evolutionary relationship of the divergent genes, compared to human hemoglobin beta.
5. Make another cladogram with the second closest *divergent* sequence to human hemoglobin beta as the reference. This will probably be gorillas or orangutans.

TABLE 11.1
Accession Numbers of Globin Sequences

Protein and Species	Accession Number
Hemoglobin beta, human (*Homo sapiens*)	NP_000509
Hemoglobin alpha, human (*Homo sapiens*)	NP_000508
Hemoglobin beta, chimpanzee (*Pan troglodytes*)	XP_508242
Hemoglobin beta, bonobo (*Pan paniscus*)	XP_003819077
Hemoglobin beta, orangutan (*Pongo abelii*)	XP_002822173
Hemoglobin beta, gorilla (*Gorilla gorilla gorilla*)	XP_018891709
Hemoglobin beta, spider monkey (*Ateles paniscus*)	Q6WN22
Hemoglobin beta, platypus (*Ornithorhynchus anatinus*)	P02111
Hemoglobin beta, goose (*Anser indicus*)	ACT81104
Globin-like isoform, snail (*Pomacea canaliculata*)	XP_025105346
Hemocyanin 1, squid (*Todarodes pacificus*)	BAS69907
Hemocyanin 2, squid (*Todarodes pacificus*)	BAS69906

Note: The accession number for the sequences in this table can be used to access the sequences of the listed proteins through the National Center for Biotechnology Information or else through the Swiss Protein Databank.

6. Make a third cladogram using the third closest divergent sequence as the reference.
7. Platypi diverged from the evolutionary line that produced humans 166,000,000 years ago, according to the fossil record (Rowe et al., 2008) and other genetic evidence (Warren et al., 2008). Use this fact with your first cladogram to determine the gap of evolutionary time that each gene separated from human hemoglobin beta.

Next, you should repeat your analysis, but this time compare hemoglobin alpha sequences in order to generate your cladograms. You will notice that no sequence from spider monkeys is available, because the sequence of hemoglobin alpha out of this organism has not at this time been determined.

Protein and Species	Accession Number
Hemoglobin beta, human (*Homo sapiens*)	NP_000509
Hemoglobin alpha, human (*Homo sapiens*)	NP_000508
Hemoglobin alpha, chimpanzee (*Pan troglodytes*)	NP_001036092
Hemoglobin alpha, bonobo (*Pan paniscus*)	XP_003809438
Hemoglobin alpha, orangutan (*Pongo abelii*)	XP_024089067
Hemoglobin alpha, gorilla (*Gorilla gorilla gorilla*)	P01923
Hemoglobin alpha, platypus (*Ornithorhynchus anatinus*)	XP_001517140
Hemoglobin alpha, goose (*Anser indicus*)	ACT80359
Globin-like isoform, snail (*Pomacea canaliculata*)	XP_025105346
Hemocyanin 1, squid (*Todarodes pacificus*)	BAS69907
Hemocyanin 2, squid (*Todarodes pacificus*)	BAS69906

1. Score human hemoglobin alpha against itself for a reference score.
2. Score all the other sequences against human hemoglobin alpha.
3. Create a cladogram to indicate the evolutionary relationship of the divergent genes, compared to human hemoglobin alpha.
4. Use the known separation in time between humans and platypi to determine the age of separation of the alpha genes.
5. Identify discrepancies in the distance in time that these genes separated, as determined from your two genetic analyses. The actual length of time is probably close to the average between the two times determined in these analyses. If you are doing the next part, determine the separation in time between humans and gorillas, orangutans, and spider monkeys.

Certain sequences of genes have been found in other species of humans recently, most notably Neanderthals, Heidelberg Man, and Denisovans. Genes located in the mitochondria have been entirely sequenced in all these species. One such gene is mitochondrial cytochrome b. Your previous two studies have given a length of time that orangutans and gorillas diverged from the *Homo sapiens* line. Use that information to determine how long ago these other three species of humans diverged from our line as well. Use your estimate of the separation time between *Homo sapiens* and spider monkeys to establish whether your conclusion is probable or not.

Cytochrome b (Mitochondrial) from Species	Accession Number
Modern Human (*Homo sapiens*)	ADT79912
Neanderthal (*Homo neaderthalis*)	AUD37535
Heidelberg Man (*Homo heidelbergensis*)	YP_008963999
Denisovan (*Denisova hominin*)	AQD17596
chimpanzee (*Pan troglodytes*)	AEQ35832
bonobo (*Pan paniscus*)	AEO20562
gorilla (*Gorilla gorilla gorilla*)	YP_002120670
orangutan (*Pongo abelii*)	NP_007847
spider monkey (*Ateles paniscus*)	AIZ03418

1. Score *Homo sapiens* cytochrome b against itself for a reference score.
2. Score all the other sequences against *Homo sapiens* cytochrome b.
3. Create a cladogram to indicate the evolutionary relationship of the divergent genes, compared to *Homo sapiens* cytochrome b.
4. Use the separation in time between *Homo sapiens* and gorillas as a benchmark to determine the separation in time of all other species in the cladogram.
5. Then use the separation in time between *Homo sapiens* and orangutans to determine the separation in time of all other species. There will be a slight discrepancy.
6. Using spider monkeys as a comparison, determine if the gorilla or the orangutan serves as a better benchmark of time, or if the true answer is between the two.

POST-LAB QUESTIONS

a. α_2/β_2 hemoglobin is alleged first to appear in species approximately 450 to 500 million years ago (Hardison, 2012). Do your data support this hypothesis?
b. Squid, octopi, clams, oysters, snails, and horseshoe crabs do not use hemoglobin as oxygen transporters. Instead, they use hemocyanin (Magnus et al., 1994). What do your results regarding hemocyanin suggest is the probable reason why these organisms do not have hemoglobin?
c. You get slightly different evolutionary distances if you use a different reference protein than human hemoglobin beta. Explain why.
d. (Extra) Most Caucasians have between 2% and 4% Neanderthal DNA, due to several crossbreeding events. This is despite the fact that *Homo sapiens* and *Homo neanderthalis* are separate species, not ancestors. Virtually no humans of African descent have any Neanderthal DNA. Denosovan DNA shows up with equal frequency in some Asiatic peoples as Neanderthal DNA.
   i. Is this a specimen of lateral gene transfer?
   ii. If Neanderthal hemoglobin beta were used as a benchmark for the first cladogram instead of Homo sapiens hemoglobin beta, would it have suggested different evolutionary times than you obtained? Why or why not?

REFERENCES

Hardison, R. C. Evolution of hemoglobin and its genes. *Cold Spring Harbor Perspectives in Medicine*, *2*(12), (2012) a011627.

Henikoff, S., and Henikoff, J. G. Amino acid substitution matrices from protein blocks. *Proceedings of the National Academy of Sciences*, *89*(22), (1992) 10915–10919.

Magnus, K. A., Ton-That, Hoa, and Carpenter, J. E. Recent structural work on the oxygen transport protein hemocyanin. *Chemical Reviews*, *94*(3), (1994) 727–735.

Rowe, T., Rich, T.H., Vickers-Rich, P., Springer, M., Woodborne, M.O. The oldest platypus and its bearing on divergence timing of the platypus and echidna clades. *Proceedings of the National Academy of Sciences*, *105*(4), (2008) 1238–1242.

Warren, W. C., et al. Genome analysis of the platypus reveals unique signatures of evolution. *Nature*, *453*(7192), (2008) 175.

12 Growing Crystals of Hemoglobin

One of the most important advances in the fields of biochemistry and medicine has been the development of the field of structural biology and biochemistry. Determining the structure of macromolecules has allowed for rational drug design, improved phylogenetic relationships, improved purification and chemical modification techniques, and is largely responsible for the creation of nanotechnology. The use of X-ray bombardment on crystals of molecules and the analysis of structures by the diffraction patterns they generated was developed before its use in biochemistry. The solution of the structures of calcium carbonate, sodium chloride, and even some simple organic molecules had already been achieved by this technique, before it was ever applied to biological molecules. In the 1950s, the brilliant chemist Max Perutz developed the method of molecular replacement strategies and enabled the solution of much larger structures, using diffraction patterns created by crystals of these large molecules. In rapid order, he solved the structures of myoglobin and hemoglobin, both of which were immensely larger than the largest structure that had ever been solved to that date. His technique was soon used to solve structures of many proteins, then helped reveal the structures of DNA and RNA. He even managed to show that macromolecules formed different structures if they bound to specific ligands, creating the first rational basis for explaining macromolecular communication. For his work, he and John Kendrew shared the Nobel Prize in 1965. The structure of myoglobin is shown in Figure 12.1.

Other strategies to solve macromolecular structures have also been developed, including various NMR techniques. NOESY and COSY can be used to reveal not

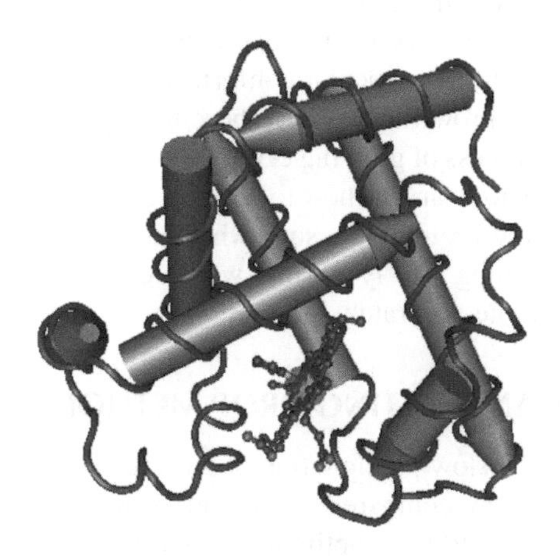

FIGURE 12.1 Myoglobin.

only the structures of macromolecules, but also the vibrational patterns and proton exchange frequencies within the structure. On the other hand, they are much more limited in size, and cannot work with large proteins given the current limits of the technology. Electron microscopy also can reveal some protein structures, but only for very large protein complexes. In addition, the atomic resolution of that technique is worse than other techniques. There are other techniques as well, but X-ray crystallography remains the most reliable workhorse in the field of structural biology, at present, offering the best resolution structures of the largest and smallest proteins with the most reliable results. It has its limitations, but it is not within the scope of a few paragraphs to spell out their advantages and disadvantages. Rather, let us start by accepting that it is an important and valuable technique, which should be known and used by the average biochemist.

Of course, if it is so important to use this technique, one must be able to generate large, high-quality crystals of the macromolecules to be studied. This is much harder to do than it sounds. The growth of useful protein crystals is something like an art form, something like lucky chance, and something like rigorous and detailed observation and manipulation of every condition that can stabilize and precipitate a particular protein. A protein that is very flexible will not crystallize without being somewhat restricted. A protein that changes its configuration as it precipitates with certain salts has to be precipitated with other ones, or it just forms amorphous crystals. Some proteins will never form crystals in the presence of oxygen, or their crystals will shatter as soon as they encounter it. There is a host of things that can go wrong. Therefore, the diligent researcher has to be somewhat persistent, but also know the principles by which the protein can be induced to form stable crystals. Furthermore, he or she needs to try many different combinations of parameters and carefully examine the results in order to refine the conditions for crystal growth.

This laboratory will help show the process of determining the conditions to grow crystals, in a case where the conditions that actually work have been determined fairly well. Hemoglobin, as has been mentioned, was one of the first proteins to be crystallized and to have its structure solved. In the decades since, improved crystal quality has led to ever better crystals of this protein in different conditions. Ligands have been added that stabilize particular conformations. The salt and pH conditions that lead to high-quality growth have been thoroughly studied. It is because of the known solution to the process of growing crystals of this protein that it makes a good educational opportunity for learning how to develop growth conditions. Additionally, since hemoglobin is a red protein, it is somewhat easier to spot crystals of it than it would be to spot something like lysozyme, whose crystal growth conditions have also been determined quite accurately.

HANGING DROP AND SITTING DROP METHODS

Any crystal forms by the slow precipitation of a chemical from solution. If the rate of precipitation is very low compared to the rate of molecular tumbling, then the molecule should have time to pack optimally with other molecules of its own kind, in regular repeating patterns. So long as precipitation does not occur before optimal packing occurs, then crystals of some type can grow. The slow precipitation can be achieved by slow salting out.

Salting out has been described previously, because it is used in protein purifications extensively. An ionic compound such as ammonium sulfate or sodium chloride attracts water to its ions better than the surface of the protein does. If the concentration of the salt is raised high enough then there are not enough water molecules to make a solvent shell around the protein, and it comes out of solution. Diffused ions such as ammonium or sulfate will be less disruptive of the protein structure than ones where the charge is more concentrated, such as sodium or chloride. Many proteins are robust enough to withstand the charge-concentrated ions, though, which allows for their use as precipitants in crystal-growth work. In the case of crystal growth, the protein is put into conditions where the precipitant concentration is *slightly below* the amount required to precipitate the protein. Such a condition can be achieved in the following manner: a reservoir of a solution whose precipitant concentration is high enough to precipitate the protein is put into a small space; then, within that space, but held separate by some means, a tiny amount of protein in solution is positioned somewhere, and an equally tiny amount of the precipitant solution from the reservoir is added to it; if equal volumes are used, then the concentration of the precipitant is cut in half, and therefore is probably less than what is required to precipitate the protein.

The position of the drop determines the method. In some methods, the drop of protein is put onto a glass or plastic cover slip, which is placed over the reservoir of precipitant such that the protein drop hangs over the reservoir. This is the "Hanging Drop" method, and the drop stays suspended on the cover slip because its weight is less than the force of friction that holds it onto the slip. The seal between the cover slip and the reservoir must be airtight, but that is easily accomplished by using vacuum grease around the top edge of the reservoir container. This method is limited in the size of crystals that can be grown, because the drop of protein can never be too large, lest it simply fall down into the precipitant solution. It is easy to do, however, and generally, the materials required are not very costly. Larger drops and larger crystals can be grown using the "Sitting Drop" method. In this method, the reservoir chamber has a small pillar at the middle, the top of which is flat or slightly concave and which remains above the surface of the precipitant solution. As before, an amount of protein solution and an equal amount of precipitant solution are mixed, cutting the precipitant concentration in half. This time, the drop is placed sitting at the top of the pillar, where it is kept out of the precipitant solution but held in place by gravity instead of by friction to the cover slip. A cover slip is still used along with vacuum grease, but this time it only serves the purpose of making the space within the reservoir chamber airtight. Both methods work well, with the latter often working somewhat better, but costing more and requiring more specialized equipment. The methods are illustrated in Figure 12.2.

After setting up conditions within multiple reservoirs for crystals to grow, the researcher must wait while *osmosis* begins the process of increasing the precipitant concentration. Most general chemistry students are taught that osmosis is the flow of water across a semipermeable membrane from conditions of low-solute concentration to conditions of high-solute concentration. This is an entirely accurate definition,

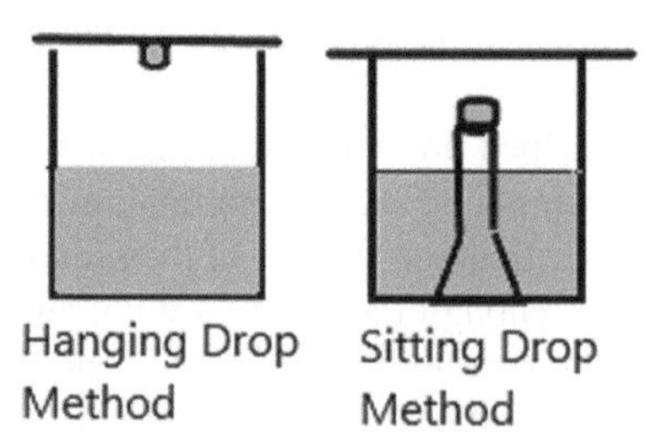

FIGURE 12.2 The hanging drop method compared to the sitting drop method. Only the placement of the droplet of protein mixed with the well contents differs.

but often students are left with the idea that the only semipermeable membranes in this world are sausage casings or cell membranes. The fact is that ANY barrier that will allow water to pass from one place to another while restricting the movement of other solutes will serve as a semipermeable membrane. Water has a vapor pressure. This means that at the surface of any aqueous solution, water vaporizes slightly, and also recondenses back into solution. Vapor pressure decreases whenever the concentration of solute is higher, regardless of the identity of the solute. This means that the equilibrium of water vaporization and condensation is shifted slightly, such that it condenses somewhat faster in high-concentration conditions than in low-concentration conditions. The drop that has the protein and precipitant has approximately half the concentration of the pure precipitant solution itself. This means that water from the protein/precipitant drop is vaporizing faster than the water in the pure precipitant solution, and condensing more slowly. If the water is all trapped into the airspace of the reservoir by the greased cover slip, then as an equilibrium phenomenon, water in liquid form from the drop has to decrease in quantity, and be transferred to the precipitant solution itself. Since the drop has so much smaller volume than the solution below it, its volume must change proportionately more. By the time the drop has decreased in volume to half its original size, the concentration of precipitant within that drop must be as high as the concentration of precipitant in the solution below. Since that concentration was high enough to precipitate the protein, then the protein will come out of solution slowly as the volume of the drop decreases. Le Chatelier's Principle accurately predicts that this must happen.

However, will it be a good crystal when it comes out of solution? It will depend upon how quickly the concentration of the drop is changing. If the precipitant concentration in the reservoir is much higher than what is required to precipitate the protein, then osmosis will occur quickly. In this case, the protein precipitates as an amorphous glass, rather than a crystal. On the other hand, if the precipitant concentration is only slightly higher than what is required to induce precipitation, osmosis will occur more slowly, giving more time for the protein to pack well as it comes out of solution. After that, getting a good crystal requires that new nucleation events in the crystal pattern not occur on some pattern that is already growing. If they did, then the resulting crystal would be aesthetically attractive but functionally useless, much as a snowflake can reveal less about the molecular structure of ice than a large and pure ice cube. This step requires control over changing environmental conditions other than just precipitation, often by adding ligands that prevent the protein

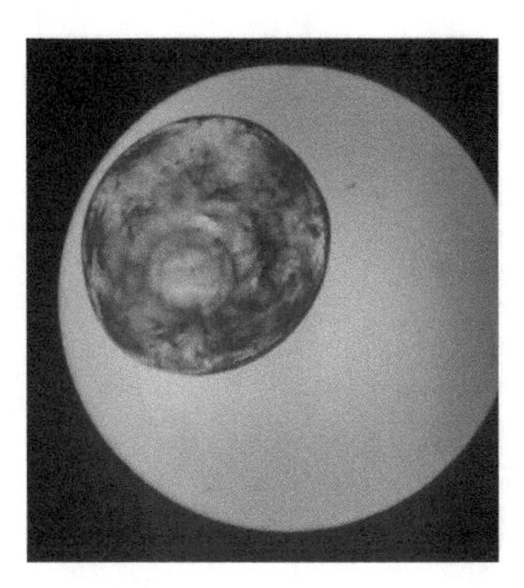

FIGURE 12.3 A glass of hemoglobin and ammonium sulfate, which is useless for determining crystal structures.

from moving very much as it precipitates. An image of hemoglobin in a glass of ammonium sulfate is shown in Figure 12.3, and is not what a scientist is attempting to obtain.

REFINING CRYSTAL GROWTH

A researcher tries a range of conditions when growing crystals. Generally, a crystal plate with multiple wells is used. This allows the researcher to vary two conditions against each other, such as precipitant concentration and pH, or two different precipitants against each other, or even ligand concentration versus some other parameter to induce slow precipitation. Every column on the plate is constant for one condition, and every row is constant for some other condition. Thus, a small 4×6 well plate, which could fit in your hand, allows for 24 different conditions of two ligands to be tested against each other.

Then, after initial crystals are grown and evaluated for size and shape – larger, more regular crystals are always better – the same conditions are tried again, but more refined. For example, suppose that in a first screen, the pH was varied from 6.0 to 7.0 in increments of 0.2 pH units, ammonium sulfate was varied from 30% saturation to 70% saturation in increments of 10%, and it was found that some crystals formed between pH 6.4 and 6.6 if the ammonium sulfate was between 40% and 50% saturation. In a second screen, the researcher would vary the pH from 6.4 to 6.6 in increments of 0.03 pH units and the ammonium sulfate from 40% to 50% in increments of 2.5%. The second screen would probably lead to even larger and more regular crystals, and would help determine more precisely the optimal conditions for crystal growth of the protein, even as other parameters were being studied at the same time. Furthermore, the information from one screen could well inform how to set up a different one. For example, if you were to find the conditions mentioned earlier, but you wanted to test whether the ligand NADH promoted crystal growth at different ammonium sulfate concentrations, you might vary the ammonium sulfate between 40% and 50%, as mentioned, and the NADH at 1 to 4 mM, but you would also set the pH at a constant 6.5 in all wells. You know that this pH is close to a

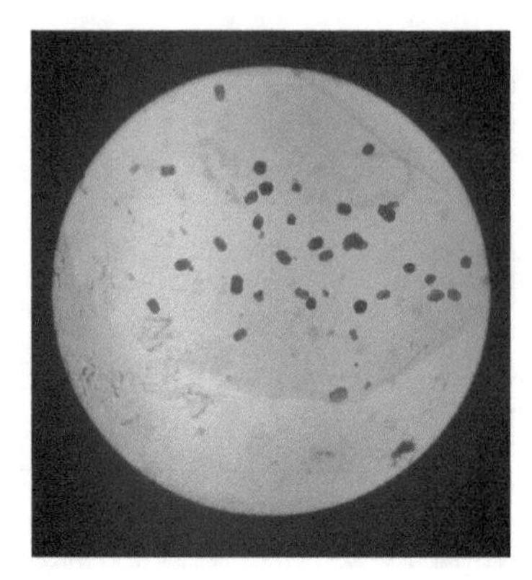

FIGURE 12.4 Microcrystals of hemoglobin look like tiny gemstones, and only form in conditions close to the ideal ones for maximal crystal growth.

value that is going to promote good crystals, and that it is presently your best guess for the optimum pH, so you would use that data while testing the relative effects of ammonium sulfate and NADH. Later on, you might use a constant concentration of NADH while varying the pH and ammonium sulfate again. The results you obtain from one set of conditions ought to inform you the next time you try to grow crystals. An image of microcrystals is shown in Figure 12.4.

After multiple rounds of these experiments – known as "crystal screens" – you frequently wind up with good crystals of your protein. A crystal that is at least 1 mm in length and width can be used in conjunction with X-ray diffraction to calculate a structure for your protein. You can evaluate the crystals for shape and size using a standard microscope.

OTHER ISSUES

Sometimes, the multiple rounds of crystal screens do not result in good crystals. There are several things to watch out for. It is usually the case that the precipitant itself will crystallize in some of your conditions. Ammonium sulfate makes large and beautiful clear crystals. Sometimes, such crystals can be mistaken for protein crystals. Should you solve the structure of such a crystal, you would be somewhat disappointed to discover what a small and uncomplicated molecule you had acquired with your crystal work. It is true that many protein crystals will also be clear, though other protein crystals can be brown, red, green, yellow, or iridescent. It is not a bad idea to find what images of ammonium sulfate or other precipitants look like when they crystallize. It will save you time and money if you recognize this frequent "false positive" before trying to use it.

Sadly, there are conditions where you cannot get a crystal of a particular protein to form at all. A few protein crystals are oxygen sensitive. A protein crystal of this kind can only be produced inside an oxygen-depleted environment, such as a "glove box". All the crystal screens have to be set up in the glove box. Since the crystals generally shatter as soon as they are exposed to oxygen, they also have to be evaluated microscopically, while still in the light box. Finally, any good crystals have to be preserved in an oxygen-deficient environment while brought to an X-ray source and bombarded. The technical difficulty is very great in a case like this, and the cost is much higher. Nonetheless, with the proper equipment it can be done.

The last common difficulty is that the conditions are sometimes interdependent, and so when you change one parameter, all the other ones shift, too. For example, if crystal growth for your protein is best at pH 6.5 and 42% ammonium sulfate, you might reasonably expect to set these conditions as constant when you screen NADH between 1 mM and 4 mM. However, at 1 mM, the optimal pH might actually be 6.2 and at 4 mM, the optimal pH might be 6.0, because the NADH induces a shift in the structure that changes the isoelectric point of the protein. As you can see, this is a condition when the information you acquire from one screen actually misinforms you as to the best conditions in another screen. It is even worse when five or six environmental conditions are interlinked. It is a solvable problem, but it requires closer attention to the results and more crystal screens than if one condition was not changing the optimal value for another condition. Sometimes, it makes the cost of creating high quality crystals greater than the available amount of money. In such cases, the crystal just is not created, and the structure is not solved until sometime when it can be afforded.

OBJECTIVE

The purpose of this experiment is to generate crystals of hemoglobin, using different crystal screens to optimize the parameters: (a) pH from 6.0 to 7.0 and ammonium sulfate between 35% and 60%; (b) pH from 6.0 to 7.0 and PEG from 8% to 18%; (c) ammonium sulfate between 35% and 60% and PEG from 8% to 18% at constant pH 6.7.

PROCEDURE

Materials

Three 24-well standard crystal plates. Sitting drop plates are recommended.

Protein Buffer: Combine 42 mL of 2 M NaH_2PO_4 with 58 mL 2 M K_2HPO_4, and adjust the pH to 6.7 if necessary. Dilute to 1.6 M by adjusting the total volume to 125 mL using deionized water.

A sample of pure hemoglobin 4 mg/mL in the aforementioned buffer. Add 4 mg sodium dithionite or DTT per mL immediately at the start of the experiment. Each student group requires a minimum of 150 μL.

2 M NaH_2PO_4 buffer, pH 6.0

2 M K_2HPO_4 buffer, pH 7.0

Glass slip covers

Vacuum grease

100% saturated ammonium sulfate solution (697 g per liter) in water

50% w/v polyethylene glycol (PEG, MW 4000) in 0.02 M phosphate pH 6.7
Deionized water

Flat toothpicks, label tape, markers, adjustable pipets (P1000, P100, P20 or P10) and corresponding pipet tips

N.B. The buffers, PEG solution, and ammonium sulfate solution should be prepared at least 24 hours in advance.

1. First, you must calculate the volumes of the various solutions you must use for the precipitation wells. This is a very time-consuming process, and is best done in your notebook BEFORE you come to lab. You want a total volume of 500 µL in each well, 0.8 M phosphate buffer total concentration regardless of the pH. You should make tables that look like the one shown below, where every row represents an individual well. All volumes are given in microliters.

   Notice that these calculations have been determined by the $C_1V_1 = C_2V_2$ *dilution mathematics. The stock concentration of ammonium sulfate is 100% and we want 60% as our concentration in a volume of 500 µL. Thus the volume of ammonium sulfate we use is*

$$\text{Volume}_{(NH4)_2SO_4} = 60\%/100\% * 500\ \mu L = 300\ \mu L$$

   This leaves a remaining 200 µL to be used for the buffer, taken from each of the sources. This volume is actually calculated by the dilution law, not by subtraction. It just happens to equal the value one gets by subtraction in this particular case:

$$\text{Volume}_{\text{buffer}} = 0.8\ M/2.0\ M * 500\ \mu L = 200\ \mu L$$

   The distribution of the buffer is easily determined by the weighted average between 6.0 and 7.0. At pH 6.0, it is 0% 7.0 and 100% 6.0, so the amount of pH 7.0 buffer you use is obviously 200 µL of the pH 6.0 buffer. However, at pH 6.3, it is 30% pH 7.0 and 7.0% pH 6.0, as seen in the calculation below:

$$6.3 = (0.7 * 6.0) + (0.3 * 7.0)$$

   *This reveals to us the volumes that are to be used. For pH 6.0 we use (0.7 *200 µL) or 140 µL. The remaining volume up to 500 µL is supplied by the water. It is for this reason that the buffer volume was not determined by subtraction: only the water volume is determined this way.*

   Fill out the volumes in the tables below. These are the volumes you will combine in the wells of your crystal trays.

Screen (a): pH vs Ammonium Sulfate

pH & conc label	Vol $(NH_4)_2SO_4$	Vol H_2O	Vol pH 6.0	Vol pH 7.0
A1: 6.0 & 60%	300	0	200	0
B1: 6.3 & 60%	300	0	140	60
C1: 6.7 & 60%	300	0	60	140
D1: 7.0 & 60%				
A2: 6.0 & 55%				
B2: 6.3 & 55%				
C2: 6.7 & 55%				
D2: 7.0 & 55%				
A3: 6.0 & 50%				
B3: 6.3 & 50%				
C3: 6.7 & 50%				
D3: 7.0 & 50%				
A4: 6.0 & 45%				
B4: 6.3 & 45%				
C4: 6.7 & 45%				
D4: 7.0 & 45%				
A5: 6.0 & 40%				
B5: 6.3 & 40%				
C5: 6.7 & 40%				
D5: 7.0 & 40%				
A6: 6.0 & 35%				
B6: 6.3 & 35%				
C6: 6.7 & 35%				
D6: 7.0 & 35%				

Screen (b): pH vs PEG

pH & conc label	Vol PEG	Vol H_2O	Vol pH 6.0	Vol pH 7.0
A1: 6.0 & 18%	180	120	200	0
B1: 6.3 & 18%	180	120	140	60
C1: 6.7 & 18%	180	120	60	140
D1: 7.0 & 18%				
A2: 6.0 & 16%				
B2: 6.3 & 16%				
C2: 6.7 & 16%				
D2: 7.0 & 16%				
A3: 6.0 & 14%				
B3: 6.3 & 14%				
C3: 6.7 & 14%				
D3: 7.0 & 14%				
A4: 6.0 & 12%				

(Continued)

pH & conc label	Vol PEG	Vol H_2O	Vol pH 6.0	Vol pH 7.0
B4: 6.3 & 12%				
C4: 6.7 & 12%				
D4: 7.0 & 12%				
A5: 6.0 & 10%				
B5: 6.3 & 10%				
C5: 6.7 & 10%				
D5: 7.0 & 10%				
A6: 6.0 & 8%				
B6: 6.3 & 8%				
C6: 6.7 & 8%				
D6: 7.0 & 8%				

Screen (c): Ammonium Sulfate vs PEG

conc labels	Vol PEG	Vol H_2O	Vol $(NH_4)_2SO_4$
A1: 35% & 18%	180	145	175
B1: 43% & 18%	180	105	215
C1: 52% & 18%	180	60	260
D1: 60% & 18%			
A2: 35% & 16%			
B2: 43% & 16%			
C2: 52% & 16%			
D2: 60% & 16%			
A3: 35% & 14%			
B3: 43% & 14%			
C3: 52% & 14%			
D3: 60% & 14%			
A4: 35% & 12%			
B4: 43% & 12%			
C4: 52% & 12%			
D4: 60% & 12%			
A5: 35% & 10%			
B5: 43% & 10%			
C5: 52% & 10%			
D5: 60% & 10%			
A6: 35% & 8%			
B6: 43% & 8%			
C6: 52% & 8%			
D6: 60% & 8%			

2. Prepare your protein solution using Buffer C and the hemoglobin. You will need a total of 150 μL for three trays, assuming you make no mistakes. Your instructor may provide you this protein, already in the buffer.

3. Place the appropriate volumes of buffer and precipitating agent, such that you actually create the conditions you calculated for in the first step. Stir the contents to mix each well. *You will notice that one parameter is constant for four wells (e.g. "60% ammonium sulfate" for the first four entries) and one parameter repeats every six times. This is because, in a 24-well plate, there are four rows of six columns. One value stays constant in every well of a given row or column. Thus, every well of the first column would have 60% ammonium sulfate, and every well of the first row would be pH 6.0. A similar pattern will exist in every crystal screen plate.*
4. Place a ring of grease around the upper lip of every well.
5. On a clean and dry cover slip, pipet 2 μL of the protein solution. Then pipet 2 μL of the contents of the well onto the drop of protein. Do not allow bubbles to be pipetted onto the cover slip. Invert the cover slip and place it over the same well you just pipetted from, making sure not to drop the protein into the well. If you do, get a new cover slip and try again, without worrying about the protein you just dropped into that well. Press down very gently on the cover slip in over to make an air-tight seal. *Alternatively*, you can place the volumes of protein and well solution on the upright post platform of the sitting drop well.
6. Repeat Step 5 for every well.
7. When you finish hanging a drop of protein over every well of a crystal screen plate, label the plate. Then store the plate in a clean location where it will not be disturbed. Generally, crystals grow better at room temperature than in refrigerated locations.
8. Crystals sometimes grow within five days, or you can wait for a couple of weeks.
9. (Days or weeks later) Bring your crystal screens to a light microscope. A magnification of 10× is sufficient. Using the microscope, examine each cover slip for crystals. Score a well as follows: 0 = no precipitation; 1 = amorphous glass; 2 = microcrystals of your protein (5 to 50 μm); 3 = small single crystals (less than 1 mm); 4 = large, high quality single crystals (larger than 1 mm). Draw a map of your crystal screen to determine condition ranges that give the best crystal results. The highest numerical values are the best results.
10. (Optional part) As a continuation, determine more refined conditions for new crystal screens, and set up three new crystal screens, repeating Steps 1 through 9. For example, if the best crystals grow between pH 6.3 and 6.5 with ammonium sulfate between 40% to 50%, you might refine the screen so that the pH ranges were 6.30, 6.34, 6.38, 6.42, 6.46, and 6.50, and the ammonium sulfate 40%, 43%, 47%, and 50%. Or else you might observe that the best crystals formed with PEG at 10%, so instead of using water in the pH versus ammonium sulfate screen, you add sufficient PEG to have a constant background of 10%. Compare the change in crystal size of the refined crystal screen to the original crystal screens.

PRE-LAB QUESTIONS

1. Why is the maximum concentration of ammonium sulfate in your crystal screens set at 60%? What would you have to change if you were going to make it be 70%?

2. Why does the rate of precipitation of the protein have to be slow in order to grow a crystal?

POST-LAB QUESTIONS

1. What conditions in the three screens led to the best crystal growth?

2. What were the largest, best crystals like in the first screen? Describe them by shape, size, and color.

3. What conditions would you select in order to improve growth in a refined set of screens?

4. What do you hypothesize will change with your best crystals in the refined screens?

5. Why was it necessary to mix the protein with an equal volume of the solution in the well, in order to induce precipitation?

6. Why is the ring of grease necessary in order to produce protein crystals?

13 Enzyme Inhibition

Enzymes are marvelous in their ability to catalyze reactions. Even more impressive is their ability to be attuned, so that they do not continue catalyzing reactions beyond what is needed biologically. They can be inhibited in their activity in multiple different ways.

Enzymes catalyze reactions according to the Michaelis–Menten scheme, shown in Figure 13.1.

According to this scheme, there is a single equilibrium present if no product is present. In this equilibrium, the free enzyme "E" binds the substrate "A" to produce the enzyme–substrate complex "EA". If the concentration of "A" goes infinitely high, it must bring all of the free enzyme to the "EA" complex where it can move on to the free enzyme and product "P". This allows for the maximal velocity V_{max} that has been seen in enzyme kinetics before. Likewise, the amount of substrate required to get equal quantities of free enzyme and enzyme–substrate complex is the Michaelis constant, previously dubbed "K_m" but here called "K_A" to specify that it is for the substrate "A" and not any other substrate.

$$K_A = [E][A]/[EA]$$

The rate, v_0, in the presence of only enzyme and substrate can be calculated according to the Michaelis–Menten equation:

$$v_0 = (v_{max} * [A])/(K_A + [A])$$

COMPETITIVE INHIBITION

Sometimes an enzyme binds both a substrate "A" and an inhibitor "I" to the same enzyme form, the free enzyme. The scheme of binding is shown in Figure 13.2.

There are two equilibria competing here, and the free enzyme is linked to both of them. The substrate still binds to the free enzyme as before and is described by the Michaelis constant K_A. The inhibitor also binds to the free enzyme, and is described by the inhibition equilibrium constant K_I where

$$K_I = [I][E]/[IE].$$

If the concentration of the inhibitor goes very high, there is less free enzyme to bind the substrate. This will pull the substrate equilibrium to the left, reducing the concentration of EA in order to replenish some of the concentration of E, according to Le Chatelier's Principle. Since the amount of EA available is being lowered, it reduces the amount of catalysis that occurs, thus inhibiting the reaction. However,

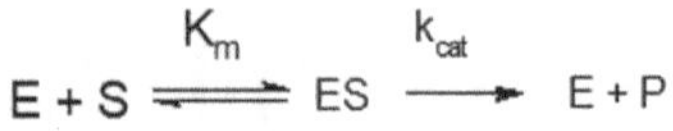

$$E + S \underset{}{\overset{K_m}{\rightleftharpoons}} ES \xrightarrow{k_{cat}} E + P$$

FIGURE 13.1 The Michaelis–Menten kinetic scheme for enzymes and other catalysts.

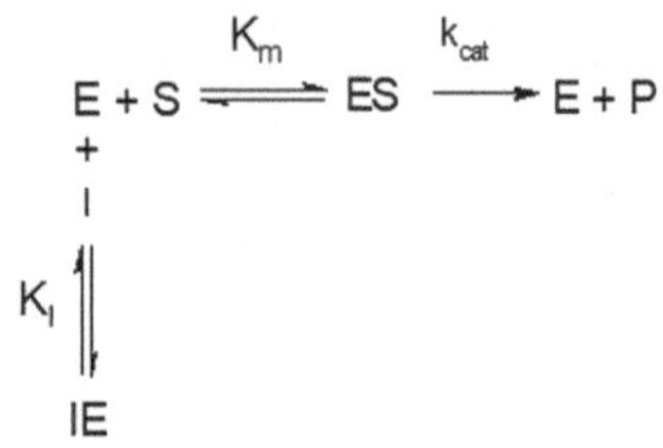

$$E + S \overset{K_m}{\rightleftharpoons} ES \xrightarrow{k_{cat}} E + P$$
$$+$$
$$I$$
$$K_I \updownarrow$$
$$IE$$

FIGURE 13.2 The kinetic scheme for competitive inhibition.

if we were to raise the concentration of A extremely high, we could still shift all of the enzyme population into the EA form, allowing for the same maximal rate of catalysis: it just takes a larger amount of substrate A to get the concentration of EA to equal the concentration of A. Thus, *a competitive inhibitor will definitely raise the value of K_A but it will not affect v_{max}.*

The apparent value of K_A, K_A^{app}, will change with the concentration of inhibitor "I", and also depends upon how strongly the protein associates with the inhibitor. The more strongly it associates, the higher the apparent value of K_A will be.

$$K_A^{app} = K_A * (1 + [I]/K_I).$$

The apparent value of the Michaelis constant replaces the actual Michaelis constant in the rate equation, since only this value changes and not v_{max}.

$$v_0 = (V_{max} * [A])/(K_A^{app} + [A])$$

$$v_0 = (V_{max} * [A])/(K_A * (1 + [I]/K_I) + [A])$$

In this last equation, we can see that the relative values of will determine the rate. If [A] becomes extremely high compared to [I], K_A and K_I, then v_0 still becomes equal to v_{max}, as Le Chatelier's Principle suggests. If [I] is extremely high compared to the other values, however, then:

$$v_0 = v_{max} * [A]/[I],$$

which approaches zero as [I] goes to infinity.

The lack of any v_{max} effect is the characteristic signature of competitive inhibition. On a Lineweaver–Burke plot, as shown in Figure 13.3, the linear fits for the data in the absence and presence of inhibitor will have the same y-intercept.

Commonly, though not always, competitive inhibition occurs when the inhibitor and the substrate bind to the same location on the enzyme.

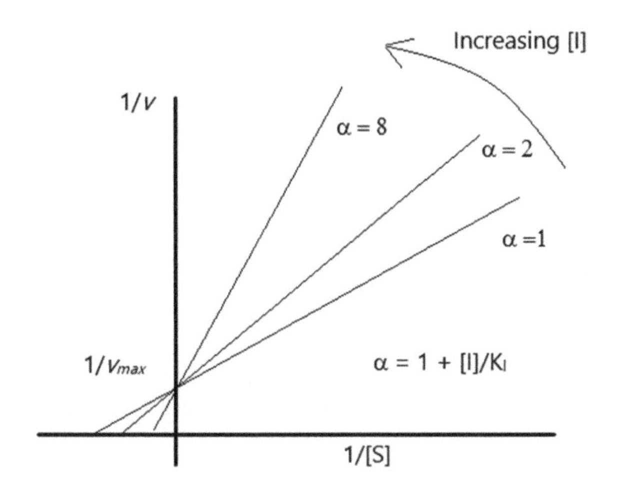

FIGURE 13.3 The Lineweaver–Burke plot of an enzyme displaying competitive inhibition. The slope grows more steep as the concentration of inhibitor increases.

UNCOMPETITIVE INHIBITION

Some inhibitors bind only to the enzyme–substrate complex, as seen in the binding scheme depicted in Figure 13.4.

There are still two competing equilibria here, but they do not share free enzyme E. In this case, if the inhibitor "I" binds, it can only do so *after* the substrate "A" has bound to the free enzyme and produced the enzyme–substrate complex "EA". However, after the inhibitor binds, it produces an inhibited complex "IEA" that cannot go on to produce product. As with competitive inhibition, it inhibits activity by reducing the available amount of enzyme substrate complex "EA".

Le Chatelier's Principle predicts a very different pattern of behavior for uncompetitive inhibitors than for competitive ones. When the concentration of substrate grows very high, it shifts the equilibrium governed by K_A further to the product and produces more EA complex. However, more EA complex shifts the equilibrium governed by K_I further towards the IEA complex. Seen in reverse, a larger concentration of inhibitor I produces more IEA complex, but that must lower the amount of available EA enzyme–substrate complex. If there is less EA complex, then the

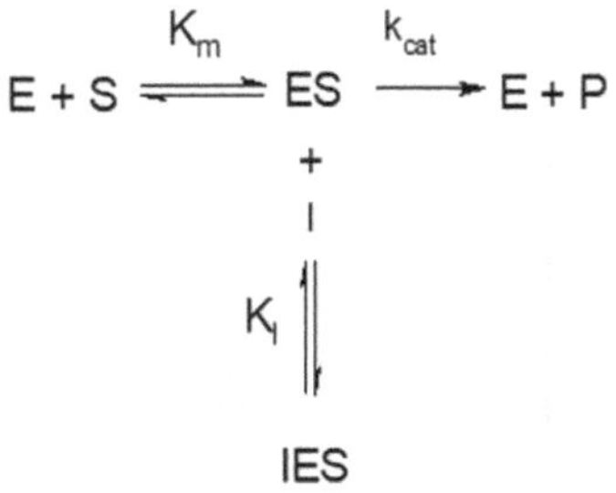

$$E + S \underset{}{\overset{K_m}{\rightleftharpoons}} ES \xrightarrow{k_{cat}} E + P$$

$$ES + I \underset{}{\overset{K_I}{\rightleftharpoons}} IES$$

FIGURE 13.4 The kinetic scheme for non-competitive inhibition.

equilibrium governed by K_A must shift to restore it, lowering the amount of free enzyme E. The effect is that a lower concentration of substrate A will be required to get equal concentrations of E and EA, since they are a smaller fraction of the total population of enzyme. If less substrate A is required, then the apparent value of the Michaelis constant, K_A^{app}, *decreases* instead of increases.

$$K_A^{app} = K_A/(1+[I]/K_I)$$

Note how this differs from the expression of K_A^{app} in competitive situations. Now a higher concentration of inhibitor shows up in the denominator, and so decreases the apparent value of the Michaelis constant.

Simultaneously, the maximal rate is changing as the concentration of inhibitor, exactly the same way as the Michaelis constant changed:

$$v_{max}^{app} = v_{max}/(1+[I]/K_I)$$

The rate will be calculated as follows:

$$v_0 = (v_{max} * [A])/(K_A + [A] * (1+[A]/K_I))$$

The increased concentration of inhibitor lowers the value of the maximal rate, and the denominator here is the same as the denominator in the relationship for K_A^{app}. This implies that there will be a constant ratio of maximal rate to Michaelis constant at any concentration of inhibitor:

$$v_{max}/K_A = v_{max}^{app}/K_A^{app} \quad \text{for any value of [I].}$$

This is the characteristic signature of uncompetitive inhibition. If a Lineweaver–Burke plot is made for the enzyme in the presence or absence of inhibitor, the linear fits of the data will be parallel, as seen in Figure 13.5.

Whenever uncompetitive inhibition is observed, the inhibitor and the substrate are not binding at the same location, structurally.

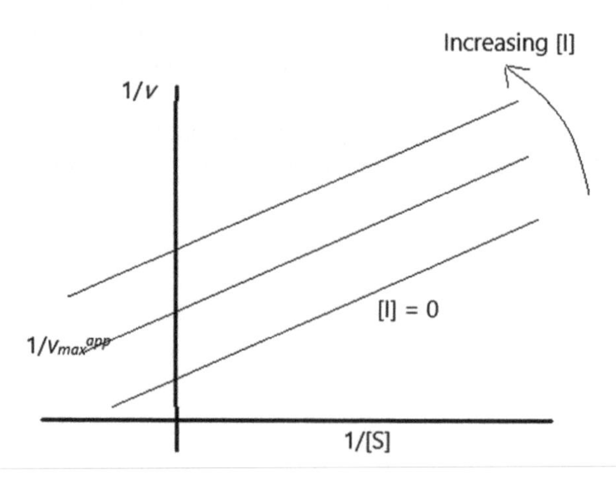

FIGURE 13.5 The Lineweaver–Burke plot of an enzyme displaying non-competitive inhibition. The slope remains parallel as the concentration of inhibitor increases.

MIXED INHIBITION

Sometimes the inhibitor is capable of binding both to the free enzyme and also to the enzyme substrate complex, as shown in the scheme in Figure 13.6

Whether the enzyme is binding to free enzyme or to complex, the result is that there is less complex "EA" to be catalyzed to product. Thus, the activity of the enzyme is inhibited. This time, there are more than two equilibrium with which to reckon. The substrate still is governed by K_A. The inhibitor, however, has two different equilibria, K_I for the inhibitor binding to the free enzyme, and K_I' for the inhibitor binding to the enzyme substrate complex.

$$K_I = [I][E]/[IE]$$

and

$$K_I' = [I][EA]/[IEA]$$

In truth, there is another equilibrium for the substrate: if the inhibitor can bind to the enzyme–substrate complex "EA", then the substrate can bind to the inhibitor–enzyme complex "IE", and is governed by an equilibrium constant K_A'

$$K_A' = [IE][A]/[IEA]$$

This is mandated by the First Law of Thermodynamics.

In certain special cases, the substrate has no effect upon how tightly the inhibitor binds, making $K_I = K_I'$. This case is known as "inhibition" and is a special case of mixed inhibition. The characteristic signature of non-competitive inhibition is that K_A^{app} always equals K_A but v_{max} decreases in the presence of inhibitor, as seen by the constant x-intercept and changing y-intercept shown in Figure 13.7.

More commonly, the inhibitory constant is affected by the presence or absence of inhibitor, as is v_{max}. The rate law therefore shows components both of competitive and uncompetitive inhibition effects.

$$v_0 = (v_{max} * [A])/\{(K_A * (1 + [I]/K_I)) + ([A] * (1 + [I]/K_I'))]$$

On a Lineweaver–Burke plot of such mixed inhibition, both the x-intercept and the y-intercept change, because the apparent values of both K_m and v_{max} are changing, as seen in Figure 13.8.

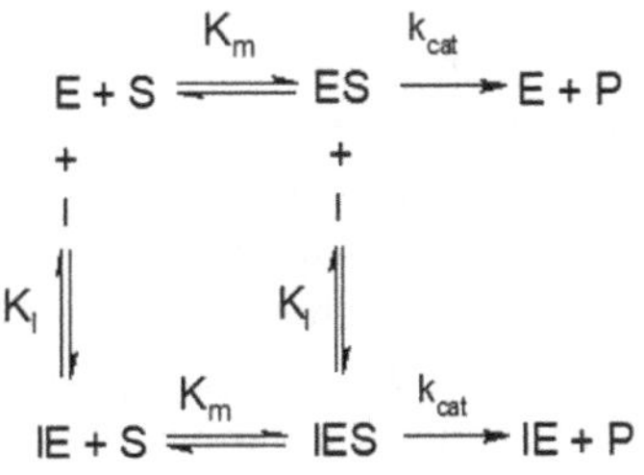

$$E + S \underset{}{\overset{K_m}{\rightleftharpoons}} ES \xrightarrow{k_{cat}} E + P$$

$$+ \qquad\qquad +$$

$$I \qquad\qquad I$$

$$K_I \updownarrow \qquad\qquad K_I \updownarrow$$

$$IE + S \underset{}{\overset{K_m}{\rightleftharpoons}} IES \xrightarrow{k_{cat}} IE + P$$

FIGURE 13.6 The kinetic scheme for mixed inhibition.

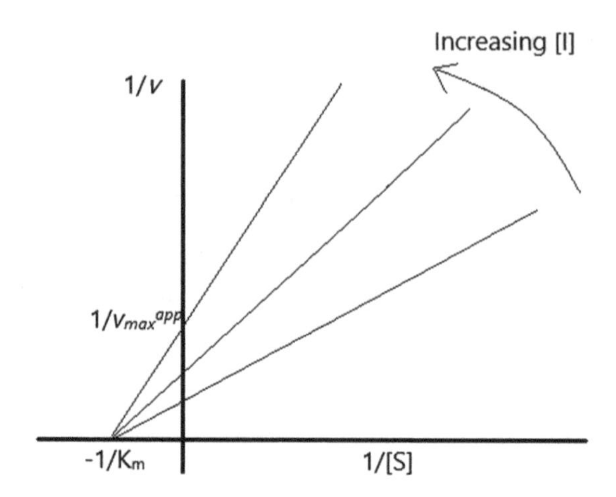

FIGURE 13.7 The Lineweaver–Burke plot of an enzyme displaying non-competitive inhibition, sometimes also called pure V-type inhibition.

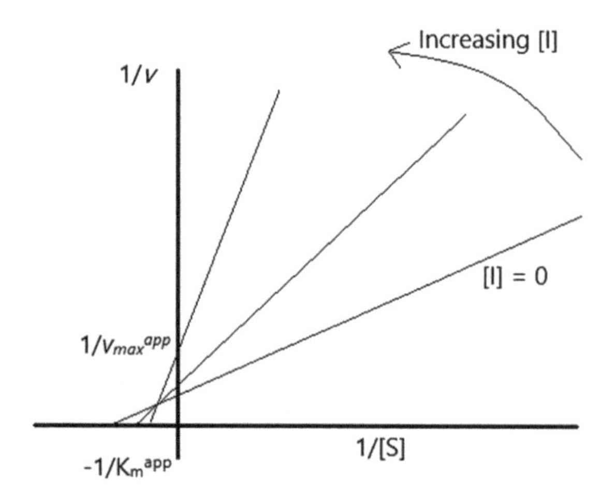

FIGURE 13.8 The Lineweaver–Burke plot of an enzyme displaying mixed inhibition, in which both the value of K_m and V_{max} change.

ACID PHOSPHATASE

Acid phosphatases are a group of enzymes that non-specifically catalyze the hydrolysis of phosphate esters. Several different types of catalysis are observed among them, including the frequent use of metal ions. One common theme among all these types of catalysis is that they are always *acid-catalyzed*. In this way, they are distinct from the equally common alkaline phosphatase group of enzymes, all of which utilize *base-catalyzed* mechanisms. Because acid phosphatases tend to have little substrate specificity, there are artificial substrates that can be used that allow the activity of acid phosphatase to be observed spectrophotometrically

$$p\text{-nitrophenylphosphate} + H_2O \rightarrow p\text{-nitrophenol} + H_2PO_4^-$$

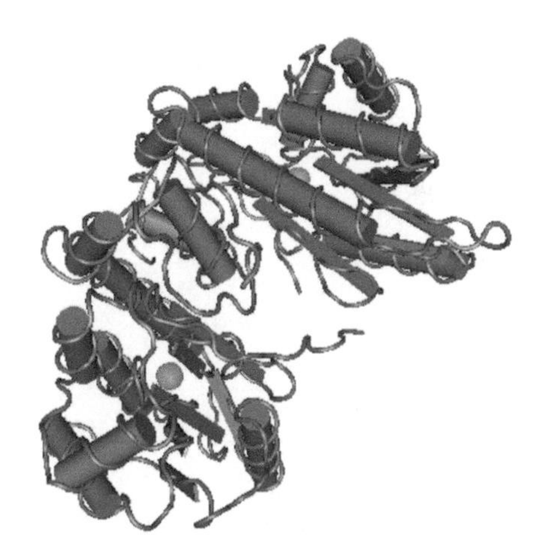

FIGURE 13.9 Alkaline phosphatase.

This reaction can be monitored at 410 nm, because p-nitrophenol absorbs strongly at that wavelength, with an extinction coefficient of 18.3 mM^{-1}.

Acid phosphatase from wheat germ uses iron catalytically, which must be able to start in the +2-oxidation state and briefly become +3 during the course of the reaction, before regaining the electron and being reduced to +2 again. If too much iron (III) is added to the solution, it will displace the catalytically used iron (II) atoms, and therefore inhibit catalysis. However, this is not displacing the PNPP ligand from the active site, so we expect non-competitive or uncompetitive inhibition from this mode of activity. The reactive iron can be seen as a sphere in the structure of acid phosphatase, shown in Figure 13.9. There are two subunits in this structure, and the iron in the upper subunit is obscured somewhat by the polypeptide backbone, but the one in the lower subunit is more visible.

Copper (II) is capable of displacing the iron altogether. Since the potential at which it oxidizes and reduces is quite different from the potential at which iron cations oxidize and reduce, the presence of copper would be expected to interfere non-competitively or uncompetitively. If the iron atom is also used to bind the PNPP ligand, then displacing it with copper should certainly trigger non-competitive behavior, but not competitive.

Phosphate itself is a product of the reaction, and as such can bind to the free enzyme. If it does so, it must be binding in the same location as the PNPP, and competing with the enzyme for the binding position. Therefore, this reaction cannot be done in a phosphate buffer, since the phosphate would naturally serve as a competitive inhibitor of the ligand. The following technique is a derivation of the alkaline phosphatase assay described by Bergmeyer et al. (1974).

OBJECTIVE

The purpose of this experiment is to determine whether copper (II), iron (III) and phosphate work as competitive inhibitors of acid phosphatase, noncompetitive inhibitors, or uncompetitive inhibitors.

Materials

Acid phosphatase from wheat germ
100 mM sodium hydroxide
p-nitrophenyl phosphate for substrate
90 mM citrate buffer pH 4.8
Spectrophotometer at 410 nm
cuvettes
Copper (II) sulfate
Iron (III) sulfate
Sodium phosphate monobasic/phosphoric acid

1. Prepare 90 mM Citrate Buffer, pH 4.8.
2. Prepare 15 mM PNPP in deionized water.
3. Prepare 1 M NaOH in deionized water.
4. Prepare 1.6 mM copper (II) sulfate in citrate buffer.
5. Prepare 20 mM iron (III) sulfate in citrate buffer.
6. Prepare 0.1 M phosphate in 90 mM citrate buffer, and adjust the pH back to 4.8.
7. Immediately before experiment begins, prepare 0.2 U/mL acid phosphatase in cold water. Store on ice. The wheat germ should be approximately 0.4 U/mg, so this means approximately 0.5 mg/mL.
8. In addition to the 15 mM PNPP stock, prepare dilutions of PNPP of 10 mM, 5 mM, 2 mM, 1 mM, 0.5 mM, 0.1 mM, and 0.05 mM, using deionized water. You will need to make serial dilutions, and after you have made dilutions, you will need at least 1.2 mL of any given concentration. Make your calculations on the table below. One of the entries is provided for you:

Dilution	Source	Volume source	Volume H_2O
15 mM	15 mM PNPP		
10 mM	15 mM PNPP		
5 mM	15 mM PNPP		
2 mM	15 mM PNPP		
1 mM	15 mM PNPP	0.134	1.866
0.5 mM	1 mM PNPP		
0.1 mM	1 mM PNPP		
0.05 mM	1 mM PNPP		

   Notice that the combined volume at the 1 mM concentration is 2 mL, but you only need 1.2 mL for other volumes. This is because the 1 mM concentration serves as the source for the next three dilutions. Also note that you will have to make four identical sets of these dilutions for the Steps 11 through 18 below.

9. Turn on a spectrophotometer and adjust it so that it is measuring absorbance at 410 nm.
10. Record the temperature.
11. In 2 mL Eppendorf tubes, as a blank for each reaction, add 0.5 mL of Citrate Buffer and 0.5 mL of the PNPP of each concentration. Mix and

allow to equilibrate for 10 minutes. After 10 minutes add 0.4 mL of NaOH and 0.1 mL of enzyme solution. Prepare the solutions in the next step at the same time.

12. In 2 mL Eppendorf tubes, determine the enzyme response at a concentration as follows: add 0.5 mL of Citrate Buffer and 0.5 mL of the PNPP of each concentration. Mix, then immediately add 0.1 mL of enzyme solution and mix again. Allow the reaction to equilibrate for 10 minutes, then add 0.4 mL of NaOH.
13. Centrifuge both the blank tubes and the experimental tubes at 12,000 rpm for one minute, in order to force any precipitate to settle. Use only the supernatant in the next step by pouring it into the cuvette.
14. At each concentration set the absorbance of the spectrophotometer to zero against the appropriate blank, then measure the absorbance of the sample. For example, you would first use the 0.05 mM PNPP blank, then measure the absorbance of the 0.05 mM PNPP test sample; then you would use the 0.1 mM PNPP blank, measure the absorbance of the 0.1 mM PNPP test sample; and so on until you finish with the 15 mM PNPP sample. Record all absorbance values.
15. The activity is calculated as follows:

$$\text{Activity(U)} = \frac{(\text{Abs}_{410\,\text{nm}} * 1.5\text{ mL} * \text{D.F})}{10\text{ min} * 18.3\text{ mM}^{-1}}$$

where the *D.F.* is the dilution factor ("1" for the conditions in this laboratory experiment).

16. To determine the inhibition with copper, prepare the blanks as described, but replace the 0.5 mL of citrate buffer with 0.5 mL of 1.6 mM copper (II) sulfate. Prepare the samples also as described above, similarly replacing the 0.5 mL of buffer with an equal volume of copper (II) sulfate. Determine the activity spectrophotometrically, just as before.
17. To determine the inhibition with iron (III), prepare the blanks as described, but replace the 0.5 mL of citrate buffer with 0.5 mL of 20 mM iron (II) sulfate. Prepare the samples also as described above, similarly replacing the 0.5 mL of buffer with an equal volume of iron (III) sulfate. Determine the activity spectrophotometrically, just as before.
18. In a similar fashion, to determine the inhibition with phosphate, prepare the blanks as described, but replace the 0.5 mL of citrate buffer with 0.5 mL of the 0.1 M phosphate solution. Prepare the samples also as described above, similarly replacing the 0.5 mL of buffer with an equal volume of 0.1 M phosphate solution. Determine the activity spectrophotometrically, just as before.
19. Construct Lineweaver–Burke plots of the activity versus the PNPP concentration, without any effector and in the presence of the three possible inhibitors.
20. Determine the values of K_m and v_{max} under all conditions.
21. Determine the type of inhibition observed, if any.

POST-LAB QUESTIONS

1. In the blank measurements, the enzyme was added after 10 minutes and after addition of NaOH, whereas in the experimental measurements, the enzyme was added right away and 10 minutes before the addition of NaOH. Why does the sequence cause the first one to serve as an appropriate blank?

2. If you observed any dark-brown precipitate form, identify whether it formed before or after the addition of NaOH. Did it form in both the blank reaction tubes, the test reaction tubes, or both? Consider the hypothesis that it is a result of the product of the enzymatic reaction: is this hypothesis consistent with what you observed? Why or why not?

REFERENCE

Bergmeyer, H.U., Gawehn, K., and Grassl, M. (1974) *Methods of enzymatic analysis.* H.U. Bergmeyer (Ed.), Volume I, 2nd ed., 495–496, New York: Academic Press.

14 Multisubstrate Kinetics

PRODUCT INHIBITION

Often, when we consider inhibition, there is a bias that the inhibitor is a substrate analog, but which will not actually undergo any catalysis to any product. Often, of course, that is true, such as when cyanide inhibits catalase, by serving as an analog to hydrogen peroxide and binding in its place at the active site. However, another molecule that certainly binds at the same location as the substrate is the product into which it is converted. Necessarily, the product must compete with the substrate for the active site. In the case of multiple substrates and multiple products, each product may compete against one or more substrates, depending upon the sequence in which each is bound and released. With two substrates resulting in two products, there are three possible scenarios, as determined by W. W. Cleland (1963a, b).

A. Ordered, Sequential Binding (see Figure 14.1): Both substrates A and B bind in order, are converted into products P and Q, which are released in order.

In this case, the substrate A and the product Q are both trying to bind to the same form of the enzyme. They will compete with one another, which means that as you increase the concentration of this product, the apparent value of K_m for substrate A will increase. However, product P and substrate B are not trying to bind to the same form of the enzyme and will not be competitive to each other. Nor, for that matter, wlll product P compete with substrate A, nor will product Q compete with substrate B. If the biochemist observes a situation where only one substrate is competitive with only one product, it indicates that the enzyme proceeds in an ordered, sequential fashion, and it also identifies which substrate binds first and which product is released last.

If the pattern of binding is equilibrium ordered and sequential, then it can be observed on Lineweaver–Burke plots. It can also be described by a modification to the Michaelis–Menten equation:

$$v = \frac{V_{max}[A][B]}{(K_{ia}K_b + K_a[B]) + \left(1 + \frac{[I]}{K_i}\right) + K_b[A] + [A][B]}$$

where K_a, K_b, and K_I are the true Michaelis constants for A, B, and the inhibitor, respectively, and K_{ia} is the apparent Michaelis constant for the inhibitor at the concentration of the substrate that is *not* being varied in the experiment.

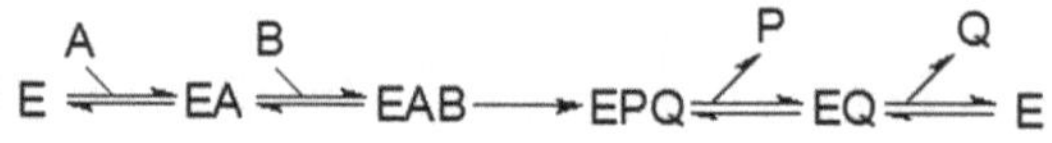

FIGURE 14.1 The binding and release scheme for ordered, sequential bimolecular kinetics.

B. Random: Substrate A and B can bind in any order and are converted into products P and Q, which can be released in any order, as diagrammed in Figure 14.2.

In this case, substrates A and B and also products P and Q are all capable of binding to the free enzyme. Because both products and substrates all bind to the same free enzyme form, E, both P and Q are competitive against both A and B. If the concentration of either is increased, they should increase the apparent K_m of either substrate, without affecting v_{max}. If this condition is observed, it often structurally corresponds to a binding pocket in which the substrates binds side-by-side. The identities of which substrate is designated "A" or "B" are not relevant, however, since either reactant could be considered "A" and either product could be considered "P".

If the substrates and products bind in a random fashion, then the initial activity can be described by the following equation:

$$v = \frac{V_{max}[A][B]}{\left(K_{ia}K_b\left(1+\frac{[I]}{K_i}\right)+K_a[B]\right)+K_b[A]+[A][B]}$$

C. Ping-Pong: As shown in Figure 14.3, substrate A binds, is catalyzed, and modifies the enzyme to form "F". Product P is released. Enzyme form "F" binds substrate B, and transfers its modification onto product Q, restoring itself to form "E" in the process.

In this case, product Q and substrate A bind to the E form of the enzyme, while product B and substrate B bind to the F form. Therefore, Q will compete with A, but will not compete with B. At the same time, P will compete with B but not A. It is not possible to designate with complete

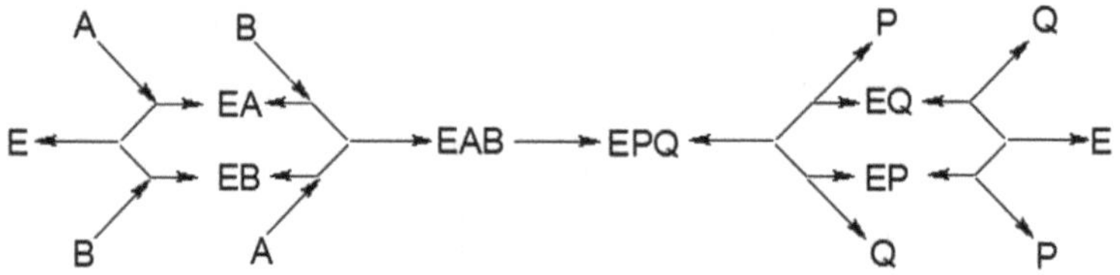

FIGURE 14.2 The binding and release scheme for random bimolecular kinetics.

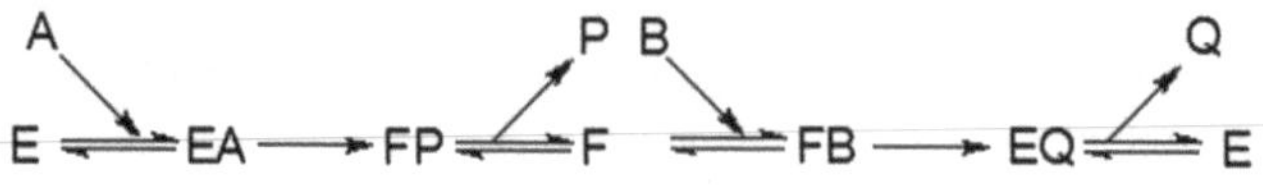

FIGURE 14.3 The binding and release scheme for ping-pong bimolecular kinetics.

certainty which substrate is the true "A" and which the true "B", but it is possible, having arbitrarily designated one of the substrates "A" to assign the identities of everything else. Whenever one product is competitive to one substrate and the other product is competitive to the other substrate, the enzyme is always operating in a ping-pong fashion. Many enzymes that transfer phosphate groups will display ping-pong behavior.

The rate equation is more complicated when ping-pong behavior is observed:

$$v = \frac{V_{max}[A][B]}{\left(K_{ia}K_b + K_a[B]\right)\left(1+\frac{[I]}{K_{i1}}\right) + K_b[A] + [A][B]\left(1+\frac{[I]}{K_{i2}}\right)}$$

where K_{i1} and K_{i2} are the two potential equilibria for binding either to the E or F forms of the enzyme. In many cases, binding affinity for one enzyme form or the other is so weak that the other equilibrium may be ignored. This is not always the case, however.

In summary, the following competition is observed

Bimolecular behavior	**P competes with …**	**Q competes with …**
Sequential, Ordered	Neither A nor B	A only
Random	Both A and B	Both A and B
Ping-Pong	B only	A only

EXPERIMENTAL DESIGN

When determining K_a, for the reaction

$$A + B \rightarrow P + Q$$

it is important that the enzyme's activity be limited only by the concentration of substrate. To this end, when determining K_a, the activity cocktail could be done with saturating concentrations of [B], while varying the concentration of [A]. Similarly, when determining K_b, the activity cocktail could have saturating concentrations of [B], while varying [A]. If you were only trying to find K_a and K_b, and nothing else, then using saturating concentrations of the other substrate is the correct strategy.

However, when determining the type of inhibition, you do not want saturating concentrations. If you were trying to find out if product P competed against A, if you made the enzyme cocktail with saturating [B], it might easily reduce the total population of enzyme to which P could bind. As you varied [A], you might not only be competing against P, but also affecting the interaction of P with B. A better strategy is to have a concentration of [B] that equals K_b through your cocktail. Therefore, when you are determining whether the product is competing with either reactant, first determine K_a and K_b and afterwards set each reactant equal to its dissociation constant in the cocktail where it is the one *not* being varied.

With certain enzymes, the ligand binds with cooperativity. This phenomenon can be recognized on Lineweaver–Burke plots by a non-linear relationship between $1/v$ and 1/[substrate]. If this occurs, you may have to determine the dissociation constant and maximal activity using the Hill equation:

$$v = \frac{V_{max}[A]^n}{\left(K_a^n + [A]^n\right)}$$

where "n" is the Hill coefficient, which must be determined by non-linear regression.

OBJECTIVE

To determine the binding order of substrates, for the enzyme L-lactate dehydrogenase.

MATERIALS

(Unless specified otherwise, all solutions to be made in deionized water.)

L-Lactate dehydrogenase, 0.1 mg/mL in phosphate buffer pH 7.0
Phosphate buffer pH 7.0
0.5 M CAPS buffer pH 10.0
6 mM NADH
450 mM L-lactate
18 mM NAD^+
7.5 mM pyruvate
Deionized water
UV/VIS spectrophotometer set to absorbance mode at 340 nm.
UV/VIS cuvettes for 1.5 mL assay solutions

1. Prepare a range of dilutions of L-lactate, at 100 μL each. The concentrations you wish to use will be 450 mM, 400 mM, 300 mM, 100 mM, 50 mM, 20 mM, 10 mM, and 5 mM. Please note that these are the concentrations you *make*, not the actual concentration of L-lactate in your cocktail. You must calculate what the final concentration of L-lactate will be, using the information provided below.
2. Prepare your assay pre-cocktail for L-LDH, which should combine 0.70 mL buffer, 0.15 mL of deionized water, and 0.10 mL NAD^+ per assay. Note that the concentrations of buffer, L-lactate, and NAD^+ are somewhat higher than in a normal assay for L-LDH. Note also that this is the **pre**-cocktail, not the complete cocktail. Since you are doing a minimum of 24 assays with this cocktail, you should multiply these values by 27, so that you will have about 10% extra cocktail volume.
3. When determining K_A and v_{max} in the absence of inhibitor, do the following: to each assay cuvette, pipet 0.95 mL of pre-cocktail, add 0.25 mL of the lactate dilution and 0.25 mL of deionized water, for a total volume of 1.45 mL. Initialize the reaction by addition of 50 μL of L-LDH, and begin monitoring the absorbance change over 60 seconds.

The assay volume will be 1.5 mL, which is the information you need to use to determine what the concentration of L-lactate in your assay. *A note of caution: this enzyme displays positive cooperativity in the binding of L-lactate in most mammalian species.*

4. Determine the K_A^{app} and v_{max}^{app} in the presence of pyruvate as follows: to each assay cuvette, pipet 0.95 mL of pre-cocktail, add 0.25 mL of the lactate dilution and 0.25 mL of pyruvate, for a total volume of 1.45 mL. Initialize the reaction by addition of 50 µL of L-LDH, and begin monitoring the absorbance change over 60 seconds. The only change between this step and the preceding one is the use of pyruvate instead of deionized water.
5. Plot the initial rate of L-LDH versus the concentration of L-lactate, both in the absence and the presence of pyruvate.
6. Calculate 1/[L-lactate] and $1/v_0$ both in the presence and absence of pyruvate. Make a Lineweaver–Burke plot. Calculate v_{max} and v_{max}^{app}. Calculate K_A and K_A^{app}. If you have cooperativity, do not use the Lineweaver–Burke plot to determine these values.
7. Identify whether the pyruvate competitively inhibits L-lactate or not.
8. Determine the K_A^{app} and v_{max}^{app} in the presence of NADH as follows: to each assay cuvette, pipet 0.95 mL of pre-cocktail, add 0.25 mL of the lactate dilution and 0.25 mL of NADH, for a total volume of 1.45 mL. The initial absorbance will be higher, so you must blank the spectrophotometer again. Initialize the reaction by addition of 50 µL of L-LDH, and begin monitoring the absorbance change over 60 seconds.
9. Plot the initial rate of L-LDH versus the concentration of L-lactate, both in the absence and the presence of NADH.
10. Calculate 1/[L-lactate] and $1/v_0$ both in the presence and absence of NADH. Make a Lineweaver–Burke plot. Calculate v_{max} and v_{max}^{app}. Calculate K_A and K_A^{app}. As mentioned above, if you observe cooperativity in L-lactate binding, do not use the Lineweaver–Burke plot to determine these values. Use non-linear regression on a plot of [L-lactate] versus v^o.
11. Identify whether the NADH competitively inhibits L-lactate or not.

In a four-hour lab time, many students will find it challenging to complete the first 11 steps and the second 11 steps within a single lab period. If teams of students are working, however, one team can perform the first 11 steps in order to determine the effect of the products on L-lactate, and a second team can perform the second 11 steps simultaneously in order to determine the effect of the products on NAD^+. By combining their results at Step 23, groups of students will be able collaboratively to determine the binding order within a short time period.

12. Prepare a range of dilutions of NAD^+, at 100 µL each. The concentrations you wish to use will be 18 mM, 10 mM, 5 mM, 2 mM, 1 mM, 0.5 mM, 0.2 mM, and 0.1 mM. Please note that these are the concentrations you *make*, not the actual concentration of L-lactate in your

cocktail. You must calculate what the final concentration of L-lactate will be, using the information provided below.

13. Prepare your assay pre-cocktail for L-LDH, which should combine 0.70 mL buffer, 0.15 mL of deionized water, and 0.10 mL L-lactate per assay. Since you are doing a minimum of 24 assays with this cocktail, you should multiply these values by 27, so that you will have about 10% extra cocktail volume.
14. When determining K_B and v_{max} in the absence of inhibitor, do the following: to each assay cuvette, pipet 0.95 mL of pre-cocktail, add 0.25 mL of the NAD^+ dilution and 0.25 mL of deionized water, for a total volume of 1.45 mL. Initialize the reaction by addition of 50 μL of L-LDH, and begin monitoring the absorbance change over 60 seconds. Note that the assay volume will be 1.5 mL.
15. Determine the K_B^{app} and v_{max}^{app} in the presence of pyruvate as follows: to each assay cuvette, pipet 0.95 mL of pre-cocktail, add 0.25 mL of the NAD^+ dilution and 0.25 mL of pyruvate, for a total volume of 1.45 mL. Initialize the reaction by addition of 50 μL of L-LDH, and begin monitoring the absorbance change over 60 seconds.
16. Plot the initial rate of L-LDH versus the concentration of NAD^+, both in the absence and the presence of pyruvate.
17. Calculate $1/[NAD^+]$ and $1/v_0$ both in the presence and absence of pyruvate. Make a Lineweaver-Burke plot. Calculate v_{max} and v_{max}^{app}. Calculate K_B and K_B^{app}.
18. Identify whether the pyruvate competitively inhibits NAD^+ or not.
19. Determine the K_B^{app} and v_{max}^{app} in the presence of NADH as follows: to each assay cuvette, pipet 0.95 mL of pre-cocktail, add 0.25 mL of the NAD^+ dilution and 0.25 mL of NADH, for a total volume of 1.45 mL. The initial absorbance will be higher, as have been mentioned before, so you must blank the spectrophotometer. Initialize the reaction by addition of 50 μL of L-LDH, and begin monitoring the absorbance change over 60 seconds.
20. Plot the initial rate of L-LDH versus the concentration of NAD^+, both in the absence and the presence of NADH.
21. Calculate $1/[NAD^+]$ and $1/v_0$ both in the presence and absence of NADH. Make a Lineweaver-Burke plot. Calculate v_{max} and v_{max}^{app}. Calculate K_B and K_B^{app}.
22. Identify whether the NADH competitively inhibits NAD^+ or not.
23. Combine the results of Steps 1 through 11 with the results of Steps 12 through 22. Determine the type of multisubstrate binding which L-lactate dehydrogenase displays, and give the order in which the substrates bind and the products release.

REFERENCES

Cleland, W. W. The kinetics of enzyme-catalyzed reactions with two or more substrates or products: I. Nomenclature and rate equations. *Biochimica et Biophysica Acta (BBA)-Specialized Section on Enzymological Subjects*, *67*, (1963a): 104–137.

Cleland, W. W. The kinetics of enzyme-catalyzed reactions with two or more substrates or products: II. Inhibition: Nomenclature and theory. *Biochimica et Biophysica Acta (BBA)-Specialized Section on Enzymological Subjects*, *67*, (1963b): 173–187.

15 Fluorescence and Denaturation

Protein stability is determined by the primary structure. The sequence of amino acids dictates what regions and types of secondary structure will exist and all the tertiary and quaternary interactions that can form. The sum total of the energy in these interactions, the charge attractions formed or lost, the gained or lost interactions of water, and the total change of entropy created by these interactions – this total amount of energy is the basis of the stability of the protein. Furthermore, the more stable the protein is, the more energy will be required to unfold the protein.

Monitoring this energy can be done in a number of ways. (1) the temperature can be increased, until the structure melts; (2) the pressure can be increased, until the structure similarly gets disrupted; (3) some denaturant chemical can be added until the energy of remaining folded is less than the energetic penalty imposed by the denaturant, and as above the structure is disrupted. All these methods result in what is referred to as a *denatured* protein: one which no longer has organized structure. All these methods also allow for quantitative measurement of the amount of energy of folding.

DENATURANTS

A number of chemicals exist that will disrupt the structure of proteins, but the two cheapest and most popular to use are urea and guanadinium chloride. Both are *chaotropic agents*, which means that they are chemicals that increase the total entropy of a system when they dissolve in water. Most solutes actually decrease the entropy in aqueous solutions, such that the entropy of the solution is less than the entropy of pure water, but for a number of reasons chaotrophic agents actually result in more disorder and therefore a higher entropy. The higher amount of entropy – and the tendency to move towards it – represents a reservoir of energy upon which other things can draw. In the case of proteins, if they are folded into an energetically stable state, it would take energy to unfold them. The reservoir of entropic energy in a solution of chaotrope would be a source of free energy to these proteins, and therefore they would tap into it and unfold. The reaction could be envisaged as follows:

$$\text{Protein}_{\text{folded}} + \text{X} \rightleftarrows \text{Protein} \bullet \text{X}_{\text{unfolded}}$$

where "X" represents the chaotropic agent. "Protein • $X_{unfolded}$" does not represent a specific interaction of the protein and the chaotropic agent, but merely the unfolded protein in the presence of the agent. It does not imply a stoichiometric relationship of any kind. The reverse of this reaction depicts the protein folding:

$$\text{Protein} \bullet X_{\text{unfolded}} \rightleftarrows \text{Protein}_{\text{folded}} + X$$

and would have the folding equilibrium constant

$$K_{\text{fold}}[\text{Protein}_{\text{folded}}] + [X]/[\text{Protein} \bullet X_{\text{unfolded}}]$$

When expressed this way, it is clear that there will be some concentration of [X] at which the amount of protein that is folded will equal the amount of protein that is unfolded. It provides an easy measurement of the value of K_{fold}, if there is any means to monitor the amount of folding. Conceptually, this allows us to determine the energy of folding based upon the concentration:

$$\Delta G_{\text{fold}} = -RT \ln K_{\text{fold}}$$

So, a researcher who carefully measures the value of K_{fold} at a given temperature will also have a measure of the stability of the protein (see Figure 15.1).

It should be noted that the free energy that is measured this way is not an absolute value of energy, but rather one proportional to other proteins only as measured using urea. Guanidinium chloride unfolds proteins at much lower concentrations, because on a molecule-to-molecule basis, it creates a much larger reservoir of entropy. The free-energy changes of one protein to another, as measured with any single denaturant, will be *proportional* to those found using any other denaturant, but they will not be identical in value.

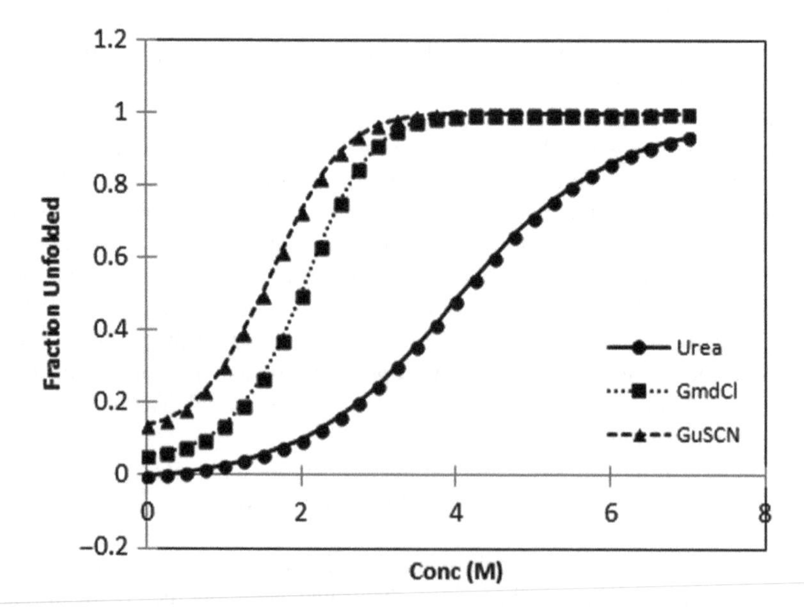

FIGURE 15.1 The stability of a protein decreases as the concentration of denaturant increases, leading to a larger fraction being unfolded.

STRUCTURAL STUDIES BY FLUORESCENCE

A number of different methods exist to study the structure of the protein, and whether it is unfolded. To treat them all is far beyond the scope of a single laboratory period, for they include activity studies, calorimetry studies of multiple different types, circular dichroisms, visible and ultraviolet spectroscopy, and of course fluorescence. It is fluorescence that we will treat upon here.

Any chemical substance can absorb light of certain wavelengths, according to Beer's Law

$$Abs = \varepsilon bc$$

where b is the pathlength of light through a sample, c is the concentration of the sample, and ε is the intrinsic absorptivity of light for that chemical. In point of fact, it is the bonds of the chemical itself that are doing the absorbing, not anything else inherent to the chemical. Consequently, in larger molecules, the large collection of bonds comprising the functional groups can be said to do the absorbing, as if they were independent molecules of the larger macromolecule that they actually comprise. In a protein, we can say that the tryptophan, tyrosine, and phenylalanine residues are doing the absorbing of ultraviolet, instead of saying that the protein itself is doing so. Conceptually, it is common to think this way in biochemistry, but chemists that work with smaller molecules tend to think of the whole molecule doing the absorbing. For our purposes, we can say that the aromatic amino acids absorb ultraviolet light well in a protein.

Notably, tryptophan absorbs ultraviolet light better than any of the other amino acid residues. It has a large aromatic structure with a heteroatom, leading to efficient absorption of a photon into its conjugated pi bonds. It also has two planes of absorbance, which are perpendicular to each other, allowing efficient absorption in two different ways. This is shown in Figure 15.2.

When it absorbs light, the energy of absorption promotes electrons to higher orbital energy levels, as seen in Figure 15.3. Each change of energy level represents a

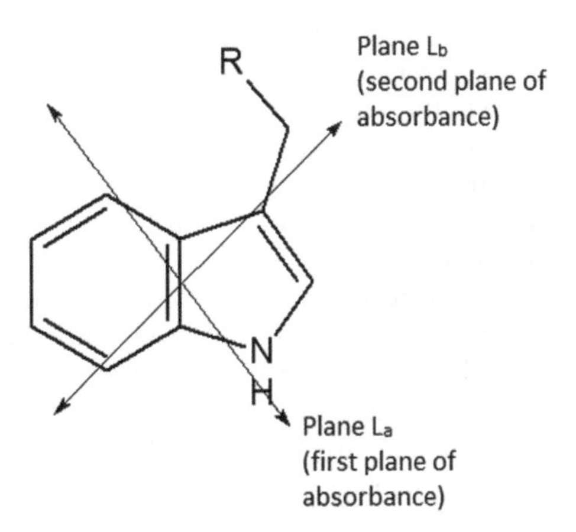

FIGURE 15.2 The indole ring of tryptophan has two planes of absorbance for light.

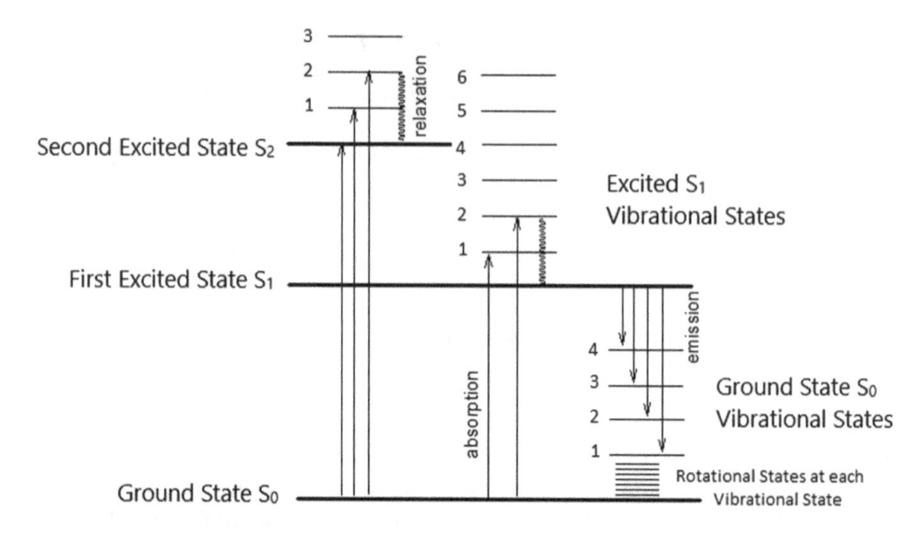

FIGURE 15.3 A photon can excite an electron from any orbital to any other, and can do so at multiple different vibrational and rotational states within that orbital. When the electron relaxes, it releases heat and sometimes emits a photon as well.

major wavelength of light, a wavelength which possesses the exact energy difference between the energy levels:

$$E = h * c/\lambda = E_{orbital\ 2} - E_{orbital\ 1}$$

Thus, initially, there are only a few discrete wavelengths that should be absorbed by the tryptophan.

Within each orbital energy level, there are different vibrational levels to which the electron can go. The different vibrational levels are variations of the same orbital, and are at slightly different energy levels, so that an electron could be at the first orbital energy level in the first vibrational mode, but another electron in a different tryptophan could be at the first orbital energy level in the second or third mode just as readily. There is just a small energy difference. When the electrons absorb the photon of light and go to the higher orbital energy level, they don't necessarily have to go to the lowest vibrational level of that orbital. They could also go to the second, third, fourth, or higher vibrational energy level, with only a small difference in energy. This means that there will be a much larger array of energy levels that represent an electron jumping from the first to the second orbital energy level. Therefore, there will not really be a few discrete wavelengths that are absorbed by the tryptophan, but rather a larger number of closely grouped ones.

To complicate the absorbance even more, within each vibrational energy level, there are a number of different rotational energy levels that also represent different energy states. An electron can jump from the first orbital energy level at any vibrational energy level with any rotational energy level up to the second orbital with any vibrational or rotational energy state as well. The number of possible wavelengths that will be absorbed is therefore enormous, and appear to be peaks of absorbances rather than discrete lines. In this way, tryptophan appears quite different from something as small as hydrogen, which has many fewer possibilities, and therefore quite discrete absorbance lines. Some wavelengths will be absorbed better, and some worse, according to frequency of electrons making the jump from one energy level to another. It will be more common for electrons to exist at the lowest possible energy state of the first orbital and go to the lowest possible energy state of the second orbital than it would be for an electron to be anywhere else or go anywhere else. The value of Ɛ comes from the frequency of electrons being in a particular starting state and going to a particular ending state.

The array of wavelengths that are absorbed and their relative values of Ɛ are referred to as the "*absorbance spectrum*" of the tryptophan.

After the electron has been absorbed, it is ultimately going to have to come back down to a less energetic state. There are a number of ways that it can do so. Most commonly, the molecule will collide with something else, and will release the energy as heat, as the electron relaxes. This almost always happens immediately after absorbance if the electron jumps to the second orbital at any other vibrational level than the first one. It relaxes to the lowest vibrational level of the second orbital, and the difference in vibrational energy is lost as heat, which gets absorbed by whatever bumped into the tryptophan. However, the electron still needs to drop its energy level back down to the first orbital, because nothing likes to be at a high energy state if it doesn't have to. If the molecule bumps into many other things, it is quite possible that the electron will decay to the first orbital again and lose its energy in a large burst of heat that goes to the other things that bumped into the tryptophan. On the other hand, if the tryptophan does not get jostled for a long-enough time, the electron will still relax back down to the first energy level, but will release energy in the form of a photon of light. This is the phenomenon of *fluorescence*.

Just as the absorbance of light can be studied by spectroscopy, the emission of light can be studied too. The light that is emitted also indicates the energy difference between the higher and lower orbital energy levels, and their various vibrational and rotational modes. This difference indicates what the environment around the tryptophan is like. The intensity of fluorescence indicates how often the tryptophan gets jostled by other molecules or functional groups while in the excited form. The wavelength that is emitted represents the energy level between the first and second states. The array of wavelengths that are emitted as fluorescence and their relative values of Ɛ are referred to as the "emission spectrum" of the tryptophan. Interestingly, the emission spectrum will always be at somewhat longer wavelengths than the absorption spectrum, because the electrons that are emitted were often excited to a higher level vibrational energy when they went to the second energy level, but they invariably relaxed to the lowest vibrational state, losing energy as heat before they decayed to the first orbital energy level by releasing a photon.

One can also study how *polarized* light is that is absorbed or emitted by tryptophan to determine how much the tryptophan moved while in the excited state, though we will not be doing so as part of this lab. It is just worth knowing that a fairly large amount of structural information can be deduced by studying the fluorescence of tryptophan.

FLUORESCENCE RESONANCE ENERGY TRANSFER (FRET)

If there is a single tryptophan in a protein, then the interpretation of the absorption spectrum, the emission spectrum, and the polarization can be fairly uncomplicated. All the energy that is fluoresced represents the changes to the environment around the tryptophan. However, the introduction of a second tryptophan can introduce complications, if they are within proximity to each other. Suppose the first tryptophan is parallel to the second tryptophan, which is stacked a short distance above it. If aromatic pi bonds are stacked in this way, the electrons of the first ring are partially resonating into the second ring, and vice versa. When the first tryptophan absorbs a photon into its aromatic pi electron rings, it promotes an electron to a higher energy level. The pi electron ring of the second tryptophan will be capable of absorbing some of the energy that was in the first ring. This means that the electron can decay by heat even without colliding with any other atoms: the atoms might have collided with the second ring, but it took away some of the energy from the first ring when it collided, via pi bond stacking. Similarly, it is possible that the second ring will take all the energy of the first one, and promote its own electron to the second orbital, while the first one relaxes back to the ground state without emitting any electron. Then the second ring has an electron that can decay by heat or else fluoresce on its own. As such, the intensity of fluorescence can be reduced immensely, and the polarization or emission spectrum that you observe might reflect both what happened to the first tryptophan and also to the second one. The signal becomes complicated. The phenomenon, diagrammed in Figure 15.4, is referred to as *fluorescence resonance energy transfer*, or "*FRET*".

However, the presence of FRET may be a blessing as well. The degree to which the electron's energy can be exchanged between one tryptophan and another depends upon how closely the aromatic rings are aligned to a parallel arrangement and also upon how far apart the rings are. If they are perpendicular or if they are too far away, you do not observe any FRET effects at all. Consequently, you can use this technique to find out whether these residues are close to each other or even the exact distance between them, if you are careful in your measurements.

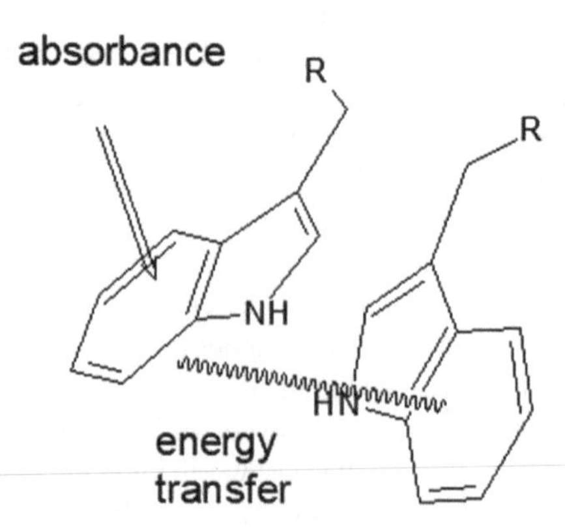

FIGURE 15.4 Fluorescence resonance energy transfer (FRET) occuring between stacked tryptophans.

FRET STUDIES FOR UNFOLDING

In the case of this laboratory, one of the proteins that is going to be used has two tryptophans in close and parallel proximity when the protein is in its native shape. For example, one of the recommended proteins is cytochrome C, a protein with a large quantity of alpha helical structure and a heme prosthetic group, in addition to the presence of multiple tryptophan residues. In its native structure, cytochrome C exhibits little fluorescence intensity. As the protein unfolds, the tryptophans move away from each other, and the fluorescence intensity grows much greater. A researcher merely monitors this change of intensity at different concentrations of urea, and determines the concentration at which the cytochrome C is halfway in between its minimum and maximum intensity values. This concentration represents the value of K_{fold}. Other proteins do not display FRET, but their unfolding can be monitored by a decrease in intensity just as easily.

EXPERIMENTAL DESIGN CONSIDERATIONS

Fluorometer Design

A fluorometer is very similar to a spectrophotometer in many ways. A spectrophotometer has three parts at a minimum and usually four: (1) a light source; (2) a monochrometer; (3) a sample cell holder; (4) usually a second monochrometer for light that passes through the sample; and (5) a detector. These are arranged in a mostly linear arrangement. At the very least, it is a linear path from the first monochrometer through the sample cell or reference cell to the detector, such that the light that reaches the detector has all passed through the sample and been partially absorbed by it.

A fluorometer has many of the same components, but the detector is at a 90° angle to the sample cell. The light that hits the detector is NOT the light that passed through the sample and was partially absorbed by it. Rather, it is the light that is emitted by the sample after the sample was struck by the energizing photons in the first place. Since this light would certainly be fainter than the light from the energizing light source, the detector must be at a 90° angle, or the detector only sees the light source, and not the light from the sample. The design difference between a spectrophotometer and a fluorometer is shown in Figure 15.5.

INNER FILTER EFFECT

When doing regular spectrophotometry, Beer's Law is said to be a linear relationship, but this is not entirely true. Absorbance is directly proportional to concentration at any given wavelength but there is a second phenomenon which is scattering, and it is geometrically proportional. If the concentration gets high enough, the light is scattered before it ever gets to the detector, resulting in no difference in light reaching the detector even at two quite different concentrations, so long as they are both high. There is a similar problem that arises with fluorescence, as seen in Figure 15.6. It is necessary that the light be detected at a 90° angle, as indicated before. This is so that the incident light of the excitation wavelength does not get to the sample. If the detector is at 90°, however, then the point of excitation in the sample being measured is only the very center of the sample compartment.

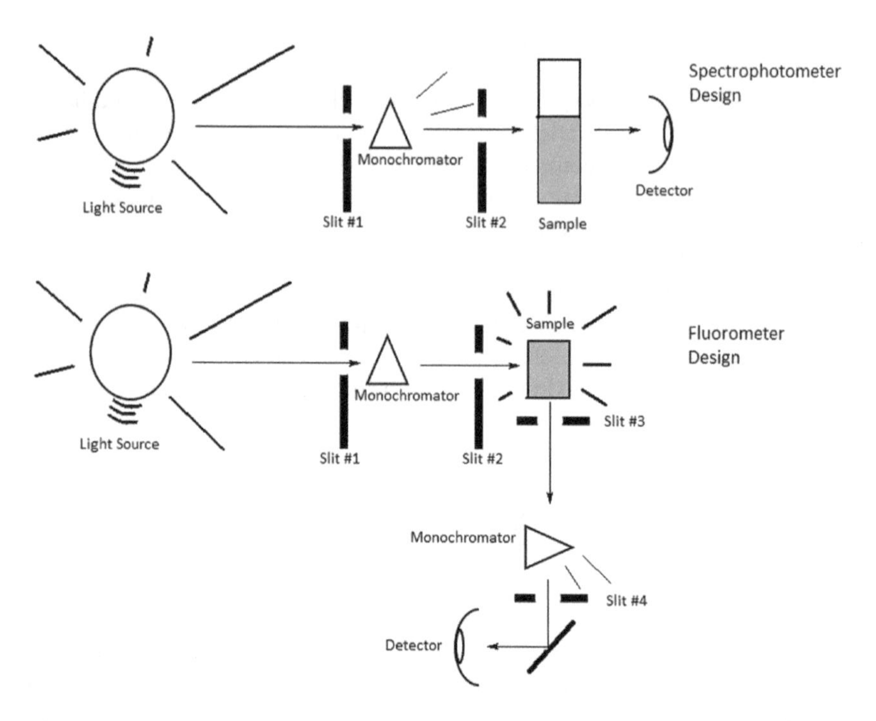

FIGURE 15.5 A comparison of the design difference between a regular spectrophotometer and a fluorometer.

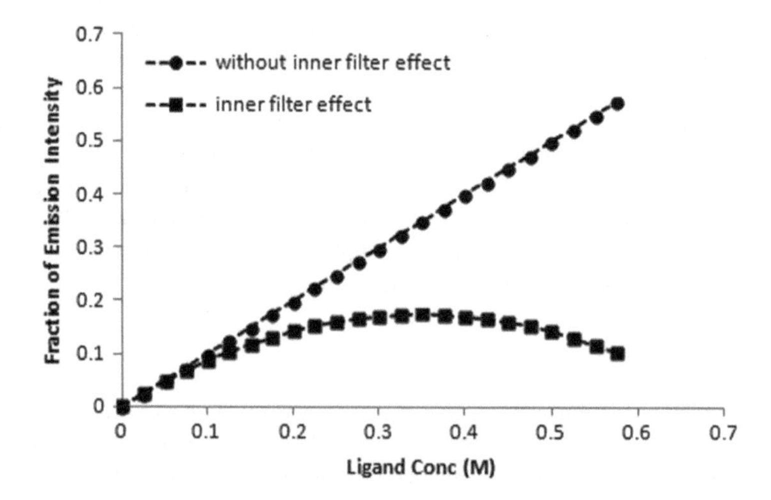

FIGURE 15.6 The inner filter effect causes a loss of signal intensity at higher concentrations, so that the observer does not see a linear response between concentration and absorbance or emission.

The light emitted by fluorescence must still pass through the sample along the path out of the sample chamber before it can get to the detector. It is assumed that the light being fluoresced is not going to be reabsorbed or scattered before it can get out of the sample chamber, but how accurate is this assumption? The emission spectrum overlaps the absorbance spectrum, so there must be some amount of reabsorbance going on. If the concentration of the sample is high, then there must be a significant amount of reabsorbance that is occurring. Additionally, if the concentration is very high, then there is a very significant amount of scattering. In fact, it might be the case that all the fluoresced light is reabsorbed or scattered before it ever gets out of the sample chamber. Therefore, FRET experiments should best be performed at protein concentrations that can be shown not to exhibit any inner filter effects.

USING AN UNFOLDING CURVE TO FIND K_{FOLD} AND ΔG_{FOLD}

In order to determine the equilibrium constant and the free energy of folding as affected by the denaturant, a graph of fluorescence intensity versus the concentration of urea must be generated. This can be done using an Excel plot or some other graphing program.

On this plot there will be one or more sigmoidal transitions, as described earlier. The intensity value will represent some folded state, which we bimodally call "folded" or "unfolded" for each transition. This is to indicate that each transition is proceeding from a more folded state to a less folded state, but does not indicate that there is a single folded or unfolded state: it is "folded" relative to its corresponding "unfolded" state, much as a weak acid is an acid relative only to its conjugate base. Each sigmoidal shape can be fit to an equation:

$$\text{fraction}_{\text{unfolded}} = \frac{e^{\left(-K_{\text{unfold}}{}^{\text{nH}} - [\text{Urea}]^{\text{nH}}\right)}}{\left(1 + \left(e^{\left(-m\left(K_{\text{unfold}}{}^{\text{nH}} - [\text{Urea}]^{\text{nH}}\right)\right)}\right)\right)}$$

where "nH" represents the Hill coefficient, a measure of cooperativity. It often is not cooperative in folding or unfolding, in which case nH = 1, but in cases of positive cooperativity, the value of nH will be greater. Values of nH equaling 1.5 to 2.5 are quite common. Higher values are sometimes observed, as are values less than 1. Each unfolding event starts at a minimal value of intensity for the folded form and results in a maximal value of intensity for the unfolded form. This means that the intensity can be fit to an equation which takes into account the starting value and the difference between the final and initial values:

$$\text{intensity} = \text{intensity}_{\text{initial}} + \left(\text{fraction}_{\text{folded}} * (\text{intensity}_{\text{final}} - \text{intensity}_{\text{final}})\right)$$

$\text{Intensity}_{\text{initial}}$ and $\text{intensity}_{\text{final}}$ refer to the intensities when the enzyme is at the start and finish of an unfolding curve, respectively.

If there is only a single transition, then the data is easy to process, and the equilibrium constant and free energy are quite simple to find. If there are multiple transitions, the free energy of each transition can be found, so long as the data reveal the upper and lower intensity plateaus sufficiently clearly. A fit of the denaturant concentration plateau transition to the intensity can be made using this equation for each transition. Then, a global fit can be made by adding the two or more intensity fits that were created. For example, the data set seen in Figure 15.7 is what is observed. Two transitions can be seen, one with K_{unfold} of 3 M, and one with K_{unfold} of 10.

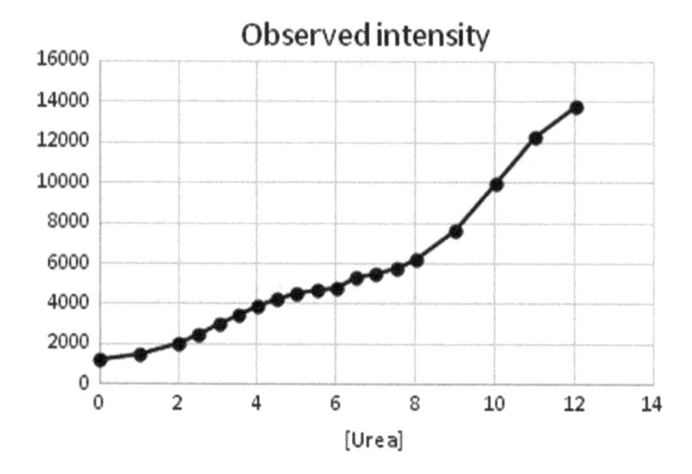

FIGURE 15.7 Two separate unfolding transitions are occurring for the protein being studied. The tryptophan fluorescence responds to each event.

The biochemist will tease out the information that led to these two curves in the following fashion:

a. The midpoint concentration of urea, K_{unfold} must be identified for each transition that is in the graph. In this case, there are two, at urea concentrations of 3 M and 10 M. They can be identified as the inflection points in the sigmoidal shapes of the curves. No cooperativity is in either of these curves, so the value of nH is set at "1".
b. For each transition, the intensities of the upper and lower plateaus are then identified. What is seen here is actually two different transitions. The one with K_{unfold} of 3 M starts at an intensity of 1000 and plateaus at an intensity of 5000. The one with K_{unfold} of 10 is obviously starting after the first one, so it must state at intensity 5000. It is harder to see where it ends, but at a guess it is around 15000. This guess is based upon the observation that the inflection point of the second transition appears to be at around 10000, which is 5000 intensity counts higher than the lower plateau: the top plateau is probably as high above the inflection point as the lower one is below it.
c. Using the values of K_{unfold}, $Intensity_{initial}$, and $Intensity_{final}$ that were identified, a curve of predicted intensities for the transitions can be generated. The two transitions, based upon their fractions folded and unfolded look as shown in Figure 15.8, and must add up to the observed total:

   In adding these transitions together, the higher intensity transition is shifted upwards by 5000, which is the final intensity of transition 1. Therefore, this value must be subtracted from the transition before adding the two curves together, seen in Figure 15.9.

   Notice that the combined transitions are now a perfect fit to the data.
d. The researcher uses non-linear regression to determine the parameters used in the transitions described earlier. The predicted combined transition fit

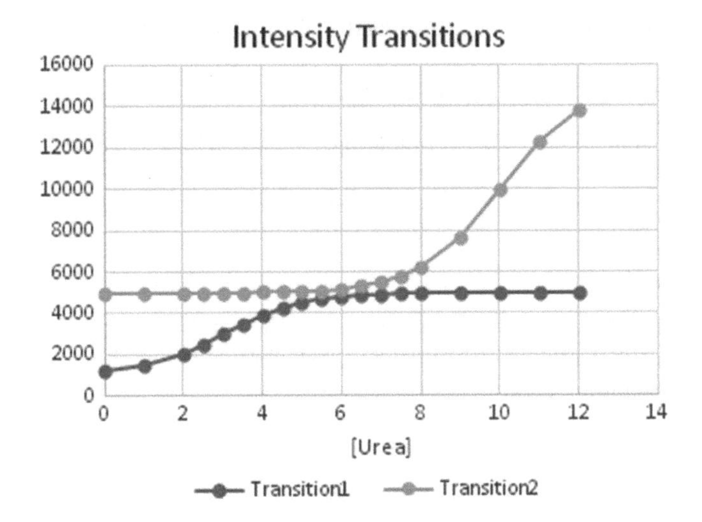

FIGURE 15.8 Two transitions are modeled separately here. The final intensity of the first transition is the initial intensity of the second one.

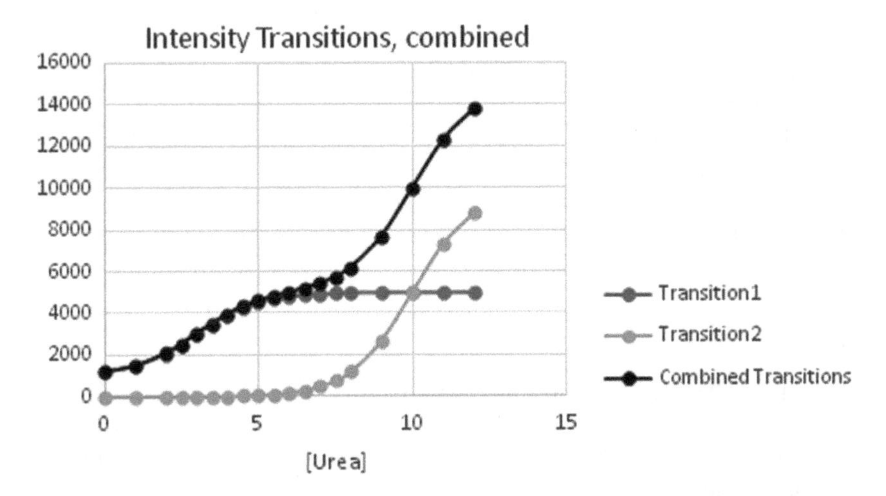

FIGURE 15.9 Two transitions are occuring in this event as well. The initial intensity of the second event has been set to zero before the two transitions are combined.

is overlaid with the data. *If it does not match the data*, then the values of K_{unfold}, $Intensity_{initial}$, and $Intensity_{final}$ for each transition must be revised until it does. In non-linear regression, the initial guesses for K_{unfold}, the initial and final values for the intensity of each transition, and the value of nH are varied one at a time, to get a fit that is closer and closer to the measured data. When the fit converges on the data, then the values for K_{unfold}, the intensities, and nH have been solved.

e. The value of K_{fold} can be determined as the reciprocal of K_{unfold}:

$$K_{fold} = (K_{unfold})^{-1}$$

and the standard free energy of folding, ΔG_{fold}, can be found by the equation

$$\Delta G_{fold} = -RT \ln K_{fold}$$

GOAL

The purpose of this experiment is to determine the equilibrium constant and standard free energy of folding of hemoglobin, cytochrome C, or albumin in the presence of the denaturant urea.

MATERIALS

You will need

- 50 mM Tris Buffer, pH 7.5
- One of the following proteins:
  - 1 mg/mL cytochrome C in 50 mM Tris Buffer, pH 7.5
  - 1 mg/mL albumin in 50 mM Tris Buffer, pH 7.5
  - 1 mg/mL hemoglobin in 50 mM Tris Buffer, pH 7.5
- Solid urea pellets
- 2 mL Eppendorf tubes
- A quartz fluorometer cuvette
- A fluorometer
- Pipets and pipet tips
- Spray bottle of deionized water

METHOD

1. Prepare 50 mL of 10 M urea in Tris buffer.
2. Using Eppendorf tubes, prepare a range of samples of protein in different urea concentrations, by creating dilutions of urea and pipetting 50 μL of the protein into each.

[Urea]$_{final}$ (M)	Vol 10 M Urea (μL)	Vol Tris Buffer (μL)	Vol Protein (μL)
0	0	1700	50
1	175	1525	50
2	350	1350	50
2.5	438	1262	50
3	525	1175	50
3.5			
4			
4.5			
5			
5.5			
6			
6.5			
7			
7.5			
8			
9			

Notice that the final volume of each sample is 1.75 mL.

$$\text{Vol 10 M Urea } (\mu\text{L}) = \text{Urea}_{\text{final}}(\text{M}) * 1750\ \mu\text{L}/10\text{ M Urea}$$

$$\text{Vol Tris buffer } (\mu\text{L}) = 1750 - (\text{Vol 10 M Urea } (\mu\text{L}) + \text{Vol protein}(\mu\text{L}))$$

3. Allow the samples to incubate for at least 15 minutes on the bench. Meanwhile, turn on the fluorometer, so that the emission lamp is warm. Results will be inconsistent if the lamp changes temperature during the experiment.
4. Bring the samples and a quartz cuvette to the fluorometer. Record the temperature.
5. Adjust the fluorometer so that the excitation wavelength is 280 nm ± 5 nm. This is the optimal wavelength for tryptophan absorbance/excitation.
6. Adjust the fluorometer so that the emission spectrum is measured from 305 nm to 400 nm.
7. Starting with the lowest concentration of urea, place the sample in the quartz cuvette, and determine the emission spectrum. Record the wavelength that emits the most ($_{\lambda max}$) and also the intensity of emission at $_{\lambda max}$. After measuring, discard the sample and rinse the cuvette with deionized water.
8. Repeat this measurement for every urea concentration.

9. Determine the total intensity within the emission spectrum range, at each concentration of urea. (Optional) Determine the wavelength of maximal emission.
10. Plot the intensity of emission versus the urea concentration.
11. Determine the value of K_{fold}.
12. Calculate the free energy of folding.
13. (Optional steps) Plot the wavelength of maximal emission versus the urea concentration. Use this plot to determine the value of K_{fold} and the standard free energy of folding. Compare whether you obtain the same values using both methods.

    If you observe two different transitions, they represent two stages of unfolding. Determine K_{fold} of the first. Generally, it represents the unfolding of the protein from the native state to the molten globule state, and the second represents the unfolding from molten globule to a completely unfolded state. The intensity often drops when it reaches this state, because of frequent collisions of water with the tryptophan.

POST-LAB QUESTIONS

1. Observe the structure of the protein you studied (hemoglobin: PDB ID = 1VWT; cytochrome C: PDB ID = 5TY3; albumin: PDB ID = 4F5S) on a molecular viewer such as Cn3D. Use the structure to predict the likely cause of the transitions you observed. For example, was FRET being observed, or just single tryptophan fluorescence? If FRET, which tryptophans were engaging in the energy transfer? At what concentration was the secondary structure lost? At what concentration did any monomers separate? Did you really observe a completely unfolded state or just a molten globule state? Justify your answer, briefly.

2. In the text, the fluorescence intensity is depicted as *increasing* when the concentration of urea increases. Under what circumstances might the intensity *decrease* upon unfolding?

3. Generally, loss of quaternary structure is observed before loss of tertiary structure. Explain why.

16 Fluorescence Studies of Ligand Binding

In previous chapters, we have studied substrate binding to enzymes, and discerned that the affinity of the substrate for the enzyme can be measured by the Michaelis constant, K_m. However, the Michaelis constant is not really the true measure of affinity, which is better expressed as the dissociation constant, K_d. It is true that in many cases, the Michaelis constant and the dissociation constant are numerically equal, but only if the catalytic rate constant, k_{cat}, is immensely less than the rate constant of substrate release, k_{-1}. In other words, the Michaelis constant only represents substrate affinity if the "steady-state hypothesis" is true. If the actual rate of catalysis is enormous, then there will be a large difference between K_m and K_d. There are many enzymes that display this behavior, including triose phosphate isomerase, superoxide dismutase, and others. Furthermore, many proteins bind ligands, but are not enzymes. Hemoglobin and myoglobin are notable examples. They have no K_m at all, and cannot be measured to have any catalytic activity in the least, and yet they do bind their ligand with a dissociation constant that exactly matches their biological needs. Even among enzymes that do not have enormous catalytic rates, there are sometimes effects caused by multiple ligands that make it impossible to equivocate K_m and K_d. If there are two substrates, sometimes one substrate inhibits the binding of the other one, causing K_m for the inhibited substrate to be far larger than K_d would be, if the substrate were binding alone. This phenomenon is observed in the enzyme *phosphofructokinase-1*, for example, in which K_m for MgATP is observed to be 49 μM in the presence of saturating Fru-6-P, but 1.7 μM when it binds alone (Johnson & Reinhart &, 1992). Similarly, a substrate can either induce cooperative binding of a second substrate, or eliminate it altogether. These varied effects can make it very difficult to determine what is really happening when any one ligand binds to a protein, if the only tool used to study the binding is the enzymatic activity.

Fortunately, binding can be studied by spectroscopic means, including fluorescence, circular dichroism, analytical ultracentrifugation, and in some cases even UV/VIS absorbance spectrophotometry. Applications of these techniques can allow a researcher to study binding independently of the enzymatic activity, revealing a more detailed picture of the enzyme's behavior. If one has access to a very powerful NMR and is working with a protein that is small enough, then it even becomes possible to study binding and protein dynamics at the same time. That is not a technique discussed in this laboratory, however.

Previous chapters have discussed the multisubstrate behavior of L-lactate dehydrogenase. The K_m values for NAD^+ and for L-lactate were measured in an experiment regarding multiligand kinetics. Herbert Fromm reported in his study of this enzyme from rabbit muscle that K_m for NAD^+ is 0.93 mM and the K_m for L-lactate is 38 mM (Fromm, 1963). Presumably, if you have measured these values yourself, you obtained similar results. What is not certain from any previous experiment you have done is whether the ligands influence each other. It is not common when measuring enzymatic activity to observe positive or negative cooperativity for L-lactate dehydrogenase. Is there no cooperativity when the ligands bind alone, however? It is widely reported in the literature that negative cooperativity of L-lactate binding is observed (Anderson, 1981; Levitzki & Koshland, 1969; Nisselbaum & Bodansky, 1961). Is the K_m itself unaltered, even if the cooperativity is altered? If so, that would imply that one ligand affects subsequent binding events differently than earlier ones. The only way to answer these questions is to measure the equilibrium constants and Hill coefficients in the presence and absence of the other ligands. A concentration versus emission intensity plot and a double reciprocal plot of a protein displaying negative cooperativity are shown in Figure 16.1.

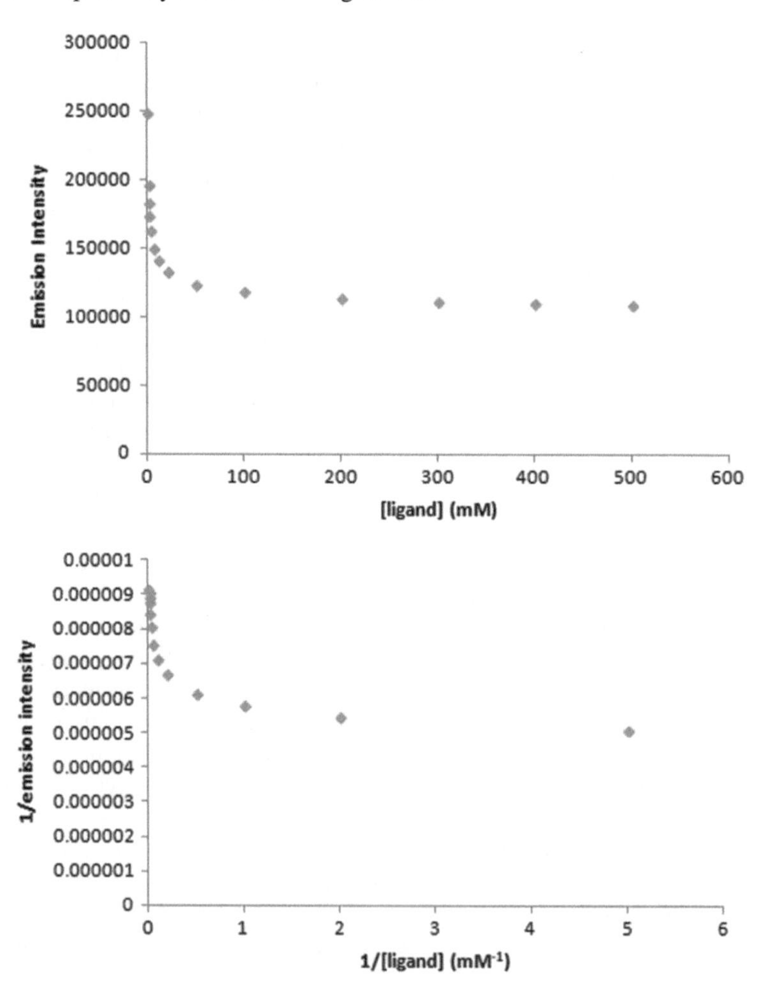

FIGURE 16.1 A protein displaying loss of intensity upon ligand binding and negative cooperativity. The Michaelis–Menten plot does not fit a hyperbolic equation and the Lineweaver–Burke plot is not linear.

In this lab, we will examine the cases when ligand binding can be studied by fluorescence. We will be able to compare the results obtained in this laboratory to known literature values (Fromm, 1963), in which K_d for NAD^+ is reported to be 0.91 mM. The basis of fluorescence has been discussed already. A loss of intensity will result from quenching of the fluorophore. A shift in the maximal wavelength of fluorescence will result from a change in the energy gap between the highest occupied molecular orbital of the fluorophore and its first excited state. If you are measuring polarization or anisotropy, a change in this property will result from a change in the restriction of movement of the fluorophore. Any of them can be used to determine binding, since the observed changes will come from the ligand binding. There are many aspects of it that can be used to study ligand binding, and we will examine them in this laboratory.

EXPERIMENTAL CONSIDERATIONS

Concentration Effects

You want to study ligand binding, not stoichiometry. This means that you must start with both the protein and the ligand concentrations below K_d. This will not be difficult with L-lactate dehydrogenase, since the enzyme itself has a somewhat weak K_d for its ligands, and it is easy to have the enzyme in lower concentration than 90 μM, the value you would need to avoid titration effects. If you try this technique with certain DNA binding proteins, however, this experiment is much less practical: certain zinc finger proteins have femtomolar dissociation constants (Kim & Pabo, 1998) and diluting a protein to less than that concentration reduces the fluorescence signal to mere noise.

Ligand Fluorescence

If your ligand fluoresces, then you must design your experiment measuring at wavelengths that do not include the fluorescence signal of the ligand itself. For example, NAD^+ does not display fluorescence, but NADH does fluoresce, emitting with a maximal peak at 460 nm. If one is studying tryptophan fluorescence, then one can take a couple of approaches to remove the NADH signal. A cutoff optical filter at 360 nm can remove the NADH signal, allowing only the fluorescence signal of tryptophan to be seen, since it is emitted at lower wavelengths than that. Another approach is to study the anisotropy (polarization) of the tryptophan emission maximum, which will occur between 300 nm and 350 nm. The anisotropy will not be affected by the NADH fluorescence. In today's experiment, the ligands that are being used will not fluoresce, because NADH is not being added.

Fitting Data to a Loss of Intensity

Fluorescence intensity can be used to monitor ligand binding, whether it increases or decreases. The data should fit mathematically much like Michaelis–Menten kinetics, or perhaps like the Hill equation, even if the intensity is decreasing. Data that fits the equivalent of the Michaelis–Menten equation will fit the following equation:

$$\text{intensity change} = (\Delta\text{int}_{\text{max}} * [A])/(K_d + [A])$$

where [A] is the ligand concentration, K_d is the dissociation constant for that ligand, and Δint_{max} is the largest change of intensity. For example if the intensity is starting out at 2,500,000 counts and maximally is going towards only 80,000 counts, the protein would have Δint_{max} of −2,420,000 counts. If cooperativity is observed, then data will fit better to a variant of the Hill equation:

$$\text{intensity change} = (\Delta int_{max} * [A]^n)/(K_d^n + [A]^n)$$

where all parameters remain the same as in the Michaelis–Menten equation, but "n" is the Hill constant, ranging between 0 and the number of active sites on the protein. A value of 1.23, for example, would indicate some positive cooperativity, and a value of 1 would indicate no cooperativity, and would revert to the Michaelis–Menten equation.

Consider the following data:

[NAD+] (mM)	Intensity
0	2491789
0.001	2440960
0.002	2356789
0.005	2199900
0.01	1945276
0.02	1617303
0.05	1044965
0.1	699112
0.2	430023
0.5	235501
1	160006
2	119911

These data would appear as in Figure 16.2 on a plot of intensity versus concentration.

To determine the value of K_d we would want to be plotting the intensity CHANGE instead of the actual intensity, so we first want to find the change from the start, essentially by subtracting 2,490,000 from each intensity value.

[NAD+] (mM)	Intensity Change
0	1789
0.001	−49040
0.002	−133211
0.005	−290100
0.01	−544724
0.02	−872697
0.05	−1445035
0.1	−1790888
0.2	−2059977
0.5	−2254499
1	−2329994
2	−2370089

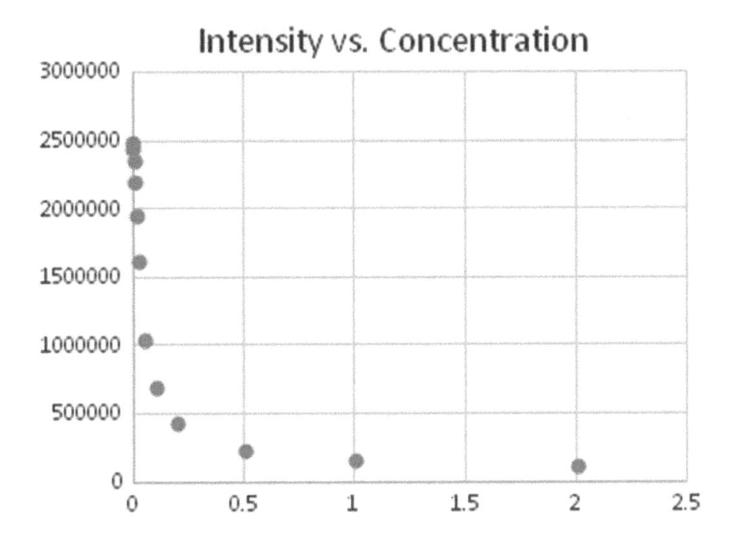

FIGURE 16.2 The data do not display cooperativity, but do show a decrease of intensity rather than an increase. It will be necessary to determine the change of intensity in order to analyze these data.

We can treat this data as we would for a Lineweaver–Burke plot and take the reciprocal of each.

1/[NAD+]	1/intensity difference
Undefined	0.000559
1000	−2E-05
500	−7.5E-06
200	−3.4E-06
100	−1.8E-06
50	−1.1E-06
20	−6.9E-07
10	−5.6E-07
5	−4.9E-07
2	−4.4E-07
1	−4.3E-07
0.5	−4.2E-07

The lowest concentration, as usual, is considered the least reliable, so, excluding the first two data points, the others points can be plotted, as shown in Figure 16.3. K_d and the maximal intensity change can be calculated as you would calculate Km and V_{max} on a Lineweaver–Burke plot.

From this plot, we can determine that

$$K_d = -1.406 \times 10^{-8} / -4.323 \times 10^{-7} = 0.032 \text{ mM}$$

$$\Delta int_{max} = 1 / -4.323 \times 10^{-7} = -2{,}313{,}000$$

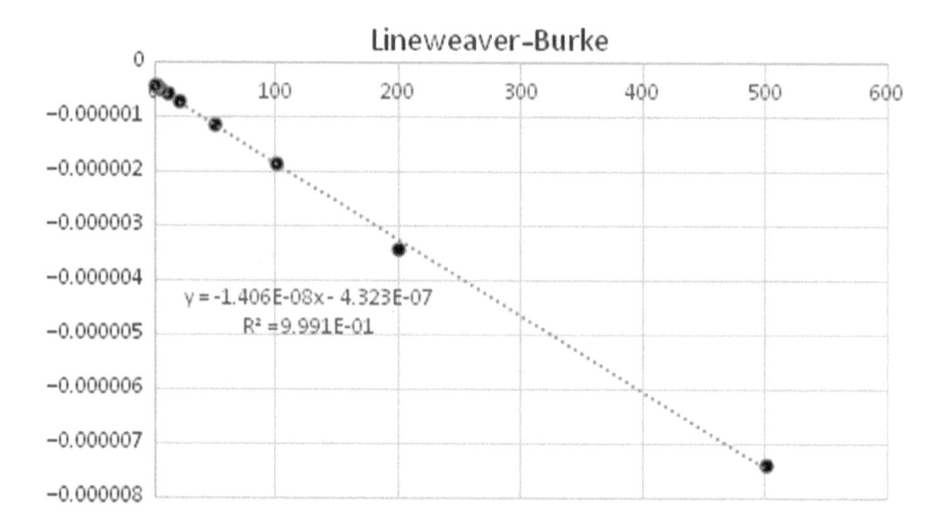

FIGURE 16.3 A Lineweaver–Burke plot of the change of intensity is linear, indicating no cooperativity. The slope and intercept can be used to determine the Michaelis constant.

We are actually only interested in the K_d value, but it is encouraging to find that our intensity change from the Lineweaver–Burk plot is so close to what we already believed it to be.

By this means, you should be able to solve for the dissociation constants of your substrates. You should also be able to do so by non-linear regression of the Hill equation or the Michaelis–Menten equation, which is not difficult in the modern age of computing.

GOAL

The purpose of this experiment is to determine the dissociation constants for L-lactate and NAD^+ to the enzyme L-lactate dehydrogenase.

MATERIALS*

L-Lactate dehydrogenase, 0.1 mg/mL in phosphate buffer pH 7.0
Phosphate buffer pH 7.0
0.2 mM L-Lactate (0.5 mL per group)
18 mM NAD^+
Deionized water
Fluorometer, with tryptophan excitation set to 280 nm and emission between 305 nm and 350 nm
Quartz cuvette, 1 cm × 1 cm

* Unless specified otherwise, all solutions to be made in deionized water.

Method

1. Turn on a fluorometer and set the excitation wavelength to 280 nm and emission spectrum detected between 305 nM and 350 nM.
2. Prepare 2 mL of L-lactate dehydrogenase, 3 mg/mL in phosphate buffer, pH 7.0. Keep this sample on ice. If you are assigned to determine K_d of both L-lactate and NAD^+, then prepare two such 2 mL enzyme samples.
3. As mentioned, you may be assigned to determine the K_d of L-lactate, of NAD^+, or both. Prepare the next step as if you were going to be assigned to do both, but your instructor may partner you with other people so that you determine only one, then compile your information with others.
4. Prepare serial solutions of L-lactate and NAD^+. You will achieve the desired concentration of each ligand by adding to the enzyme in the fluorometer, according to the table below.

Desired [Lactate] (mM)	Stock [Lactate] (mM)	Volume Added (μL)	Total Volume (mL)	Desired [NAD⁺] (mM)	Stock [NAD⁺] (mM)	Volume Added (μL)	Total Volume (mL)
0	0	0	2	0	0	0	2
0.2	4.2	0.1	2.1	0.001	0.2	0.01	2.01
0.5	6.8	0.1	2.2	0.002	0.2	0.01	2.02
1	12	0.1	2.3	0.005	0.62	0.01	2.03
2	25	0.1	2.4	0.01	1.1	0.01	2.04
5	77	0.1	2.5	0.02	2	0.01	2.05
10	135	0.1	2.6	0.05	6.2	0.01	2.06
20	280	0.1	2.7	0.1	10.5	0.01	2.07
50	860	0.1	2.8	0.2	21	0.01	2.08
100	1500	0.1	2.9	0.5	62.7	0.01	2.09
200	3100	0.1	3	1	106	0.01	2.1
500	9500	0.1	3.1	2	212	0.01	2.11

Notice that in this table it is assumed that you are adding your aliquot of the stock to the same cuvette into which you have *already* added ligand. Notice also that it is necessary to add a larger volume of L-lactate than NAD^+. This is because of the expected difference in K_d values, and the limited solubility of NAD^+ compared to L-lactate. Procedurally, it is vital that the lactate be pure L-lactate, and not a racemized mixture, because a 9.5 M stock solution of D,L-lactate cannot be made, whereas pure L-lactate is soluble up to 11.1 M at room temperature.

The amounts listed in the table mean that you will want to make 20 μL quantities of the stock solutions of NAD^+ and 150 μL quantities of lactate, unless these have been provided for you by your instructor. The volume being larger than the amount used is useful for reducing pipetting error. Determine the volumes of buffer and ligand you will have to use in order to make these serial dilutions. Part of the table is filled out for you:

Desired Stock [Lactate] (mM)	Source Conc. [Lactate]	Volume of Source (μL)	Volume of Buffer (mL)	Desired Stock [NAD⁺] (mM)	Source Conc. [NAD⁺]	Volume of Source (μL)	Volume of Buffer (mL)
0	0	N/A	N/A	0	0 mM	N/A	N/A
4.2	10 mM			0.2	0.25 mM		
6.8							
12	100 mM			0.62	2.5 mM		
25				1.1			
77				2			
135	1 M			6.2	25 mM		
280				10.5			
860				21			
1500	10 M			62.7	250 mM		
3100				106			
9500				212			

5. Bring your enzyme and ligand solutions to a fluorometer. Pour the enzyme into a clean quartz cuvette.
6. Record the temperature.
7. Determine the emission spectrum for the enzyme without ligand. Record the intensity and the wavelength of the maximal peak (λ_{max}). If your fluorometer is equipped with a polarizer, determine the anisotropy as well.
8. Add your first aliquot of ligand to the enzyme and mix it in. Determine the fluorescence intensity and λ_{max} at this wavelength, and possibly the anisotropy as well.
9. Continue adding each aliquot of ligand, so that you are increasing the concentration. At each step determine fluorescence intensity, λ_{max}, and anisotropy.
10. When you have finished all your concentrations of your ligand, clean out the cuvette with deionized water.
11. If you are determining K_d of both ligands, then pour your second enzyme solution into the clean cuvette at this point. Repeat Steps 6 through 10 above for the second ligand.
12. Turn off the fluorometer and clean your station.

Data Table

[Lactate] (mM)	Intensity	λ_{max}	Anisotropy	[NAD+] (mM)	Intensity	λ_{max}	Anisotropy
0				0			
0.2				0.001			
0.5				0.002			
1				0.005			
2				0.01			
5				0.02			
10				0.05			
20				0.1			
50				0.2			
100				0.5			
200				1			
500				2			

DATA ANALYSIS

1. Determine the value of K_d for each ligand, as described in the introduction. Expect to account for negative cooperativity in the binding of Lactate and/or NAD^+ with this enzyme.
2. If you have previously measured K_m for NAD^+ in the saturating presence of L-lactate, and a K_m for L-lactate in the saturating presence of NAD^+ (see the lab "Multisubstrate Kinetics"), then calculate the coupling constant, Q, for the enzyme (Weber, 1975).

$$Q = \frac{K_{d(NAD+)}}{K_{m(NAD+)}} = \frac{K_{d(Lactate)}}{K_{m(Lactate)}}$$

If $Q = 1$, then the ligands are neither activating nor inhibiting each other.
If $Q > 1$, then the ligands are activating each other.
If $Q < 1$, then the ligands actually inhibit each other.

Note that you only need to have K_d and K_m for one of the ligands to be able to figure out Q. You do not need it for both. If you have Q and K_m for one ligand, but cannot use fluorescence to determine K_d then calculate K_d for this ligand using the value of Q.

POST-LAB QUESTIONS

1. NAD^+ can also fluoresce. If you set the conditions of this experiment wrongly, then you must correct for your measured intensity by subtracting the intensity of from NAD^+ in solution. Find the fluorescence excitation spectrum of NAD^+ in the literature (Von Ketteler, Alexa, Dirk-Peter Herten, and Wolfgang Petrich. "Fluorescence Properties of Carba Nicotinamide Adenine Dinucleotide for Glucose Sensing." *ChemPhysChem* 13.5 (2012): 1302–1306.).

Based upon the excitation spectrum of this molecule, explain why no correction was given as part of the procedure for this lab.

2. What fluorophore is emitting fluorescence intensity in L-lactate dehydrogenase when we excite by illuminating at 280 nm?

3. The fluorescence intensity decreases upon addition of at least one of the ligands. This phenomenon is known as "quenching". Explain why this ligand can quench the intensity.

4. Sometimes, addition of a potential ligand does not cause any change in fluorescence intensity, λ_{max}, or anisotropy. A notable example is the addition of NADPH or $NADP^+$ to L-lactate dehydrogenase. If addition of a ligand does not cause any change of intensity, it does not mean that the ligand does not bind. It simply does not give any information at all. Explain why a lack of change to the fluorescent properties does not necessarily imply that there is no ligand binding.

REFERENCES

Anderson, S. R. (1981). Effects of halides on reduced nicotinamide adenine dinucleotide binding properties and catalytic activity of beef heart lactate dehydrogenase. *Biochemistry*, *20*(3), 464–467.

Fromm, H. J. (1963). Determination of dissociation constants of coenzymes and abortive ternary complexes with rabbit muscle lactate dehydrogenase from fluorescence measurements. *Journal of Biological Chemistry*, *238*(9), 2938–2944.

Johnson, J. L., & Reinhart, G. D. (1992). Magnesium-ATP and fructose 6-phosphate interactions with phosphofructokinase from Escherichia coli. *Biochemistry*, *31*(46), 11510–11518.

Kim, J. S., & Pabo, C. O. (1998). Getting a handhold on DNA: Design of poly-zinc finger proteins with femtomolar dissociation constants. *Proceedings of the National Academy of Sciences*, *95*(6), 2812–2817.

Levitzki, A., & Koshland, D. E. (1969). Negative cooperativity in regulatory enzymes. *Proceedings of the National Academy of Sciences*, *62*(4), 1121–1128.

Nisselbaum, J. S., & Bodansky, O. (1961). Purification and properties of human heart lactic dehydrogenase. *Journal of Biological Chemistry*, *236*, 323–327.

Weber, G. (1975). Energetics of ligand binding to proteins. *Advances in Protein Chemistry*. *29*, 1–83.

17 DNA Restriction Digests

DNA molecules are the principle molecule for the repository of genetic information. As such, they tend to be enormous in size, especially among eukaryotes. One of the human chromosomes, if it was extended to its full length, would stretch out over half a meter. That is not the largest DNA molecule, by a large amount: one of the chromosomes of the common daylily is over 50 meters from end to end! The amount of information available on a molecule of DNA is tremendous in size. Even bacterial chromosomes are immense in size. A plasmid is a circular piece of DNA independent of the chromosome and considered quite small by comparison. Nonetheless, plasmids can range from 1,000 to 20,000 base pairs in size, equivalent to 300 kDa to 6MDa of molecular weight. That is what is considered "small".

Because they are comparatively small, plasmids are widely used for protein engineering work in bacteria, as well as for the production of bioactive RNA molecules other than the mRNA that are used to express proteins. The genes that are on a plasmid and their architecture – *i.e.* their location, orientation, and control systems – are of great interest to the researcher.

GENE EXPRESSION

A gene is composed of a series of codons, each codon indicating that a particular amino acid must be added to the growing protein. A codon is a series of three nucelotide residues: (A) for adenine, (C) for cytosine, (G) for guanine, and (U) for uracil, as shown in Figure 17.1.

A researcher who can alter the nucleic acids of a triplet can alter specific amino acid residues in a protein. This is a common method for studying active site residues performed often by protein chemists. But the gene alone is not enough. The gene must have an ATG start codon preceded by the "Shine–Dalgarno" sequence, a six-base consensus "AGGAGG". The combination of the start codon and the Shine–Dalgarno sequence is enough to begin expression of a protein in a bacterium. Even these together, however, are not enough to produce sufficient protein for in vitro studies. To do that, the gene must be inserted into a plasmid, and the plasmid must exist in many copies within the cell. The plasmids must then be induced to overexpress the protein. The thing that many novice researchers do not see is that the DNA plasmid itself is the key to obtaining large quantities of the protein. The plasmid must be appropriate for the task assigned to it, and the location and orientation of the gene for the protein must be set very carefully.

In Figure 17.2, the plasmid known as pUC 18 is mapped. pUC 18 is a circular piece of DNA smaller than the genome. A certain base is assigned as "base 1" and numbering increases in the direction toward where a gene can be easily inserted. This will be described further in a moment. The plasmid contains more than one gene, including some form of antibiotic resistance if it is to be most useful. pUC 18 contains the

Second Letter

First Letter	U		C		A		G		Third Letter
U	UUU	Phe	UCU	Ser	UAU	Tyr	UGU	Cys	U
	UUC		UCC		UAC		UGC		C
	UUA	Leu	UCA		UAA	Stop	UGA	Stop	A
	UUG		UCG		UAG	Stop	UGG	Trp	G
C	CUU	Leu	CCU	Pro	CAU	His	CGU	Arg	U
	CUC		CCC		CAC		CGC		C
	CUA		CCA		CAA	Gln	CGA		A
	CUG		CCG		CAG		CGG		G
A	AUU	Ile	ACU	Thr	AAU	Asn	AGU	Ser	U
	AUC		ACC		AAC		AGC		C
	AUA		ACA		AAA	Lys	AGA	Arg	A
	AUG	Met	ACG		AAG		AGG		G
G	GUU	Val	GCU	Ala	GAU	Asp	GGU	Gly	U
	GUC		GCC		GAC		GGC		C
	GUA		GCA		GAA	Glu	GGA		A
	GUG		GCG		GAG		GGG		G

FIGURE 17.1 The standard translation table of codons to amino acids. There is almost no difference in codon meaning across species, despite wide evolutionary separation.

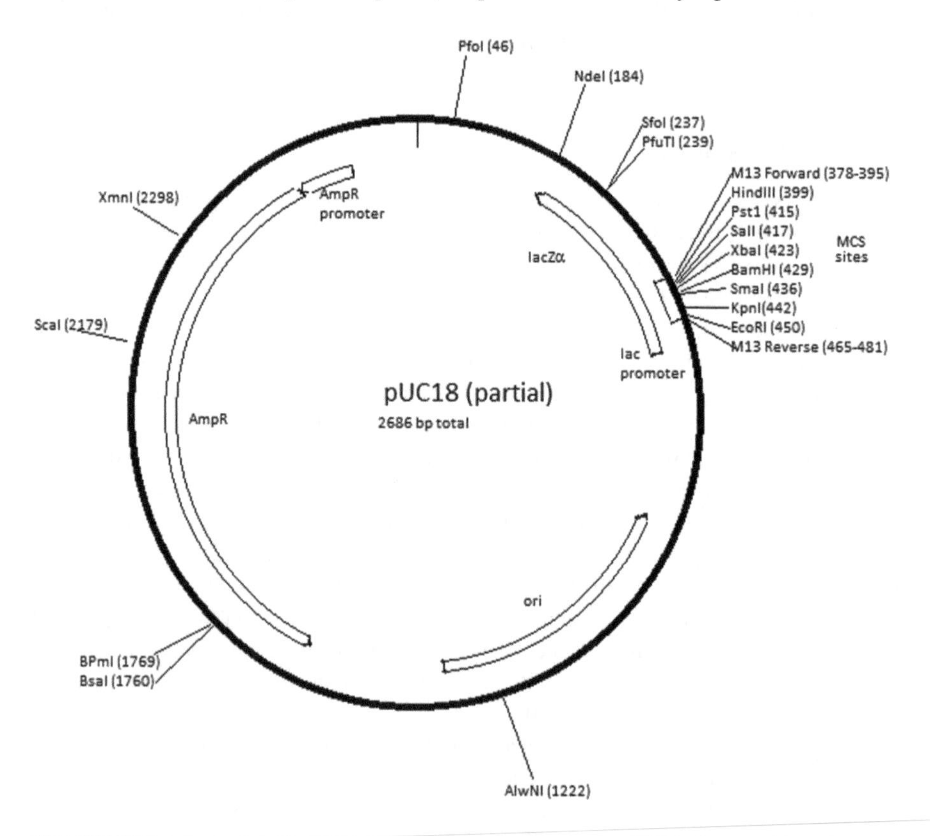

FIGURE 17.2 A partial plasmid map of the plasmid vector pUC18. Recognition sites for specific restriction endonucleases are marked according to their position around the circular DNA vector.

gene for resistance to ampicillin, which allows a researcher to kill any cell *without* the plasmid by exposing it to ampicillin. Only those bacterial cells which have been "transformed" by the plasmid would survive. Then the cell which contains the plasmid can be induced to express the protein. Different means exist in order to induce the cell to produce protein, and they will depend both on the nature of the cell and on the genetic content of the plasmid. pUC 18 has the "*lac*" operon, which allows protein expression to be induced by addition of lactose or IPTG. Other plasmids operate with other inducing agents, including chemical reagents, heat, or shaking. Some even operate without induction and just continuously induce. For the researchers using these systems, it is important that the gene they are attempting to insert into the plasmid be oriented correctly so that it will result in their protein of choice being induced and expressed. Orientation will require the judicious use of *restriction endonucleases.*

RESTRICTION ENDONUCLEASES

Bacteria have a system to counter infection by viruses. They discriminate between their own DNA and foreign DNA by recognizing certain palindromic sequences. Each bacterium has one or more *restriction endonucleases* that recognize specific palindromic sequences of DNA. If the restriction endonuclease recognizes and binds to such a sequence, it will hydrolyze the DNA strand, which will prevent replication of that DNA. For example, *Bacillus globigii* has a restriction endonuclease named "Bg1II", which recognizes the sequence 3′-AGATCT-5′, whose complement reads the same way in the opposite direction, and cuts it between the A and the G positions. The cell does not hydrolyze its own DNA, however, because it adds a methyl group to some of the base pairs in this sequence when the DNA is first replicated. The methyl group prevents the restriction endonuclease from binding and hydrolyzing. A similar strategy is employed for almost every bacterial cell that uses restriction endonucleases, of which there are thousands of varieties, each cutting at a different sequence. Each restriction enzyme is named by the first letters of the species name and then a number to designate whether it is the first, second, third, fourth, and so on that has been discovered. For example, the enzymes EcoRI, EcoRII, and EcoRIII are the first, second, and third restriction enzymes isolated from *E. coli.* HindIII is the third restriction enzyme isolated from *H. influenza.* All others are similarly named.

Scientists can use the restriction enzymes on demethylated DNA plasmids, or those purified from a source that did not methylate the palindromic sequences that would be the targets of the enzymes. Every plasmid has many sites that could be recognized by restriction enzymes, and some have multiple such sites built into a single location called the "multiple cloning site". pUC 18, for example, has a scattering of restriction sites all around its sequence, but a large collection of them have been built into the region from positions 396 to 454. It has the restriction sites for HindIII at position 399 and KpnI at position 442. If a gene of interest were isolated with a KpnI site shortly prior to its "start" codon, and a HindIII site after its "stop" codon, then exposing both the gene and the plasmid to these enzymes, allowing their cut sequences to anneal and then ligating them together with DNA ligase, would insert the gene specifically in the direction where the *lac* promoter would allow it to be induced, similarly to what is depicted in Figure 17.3.

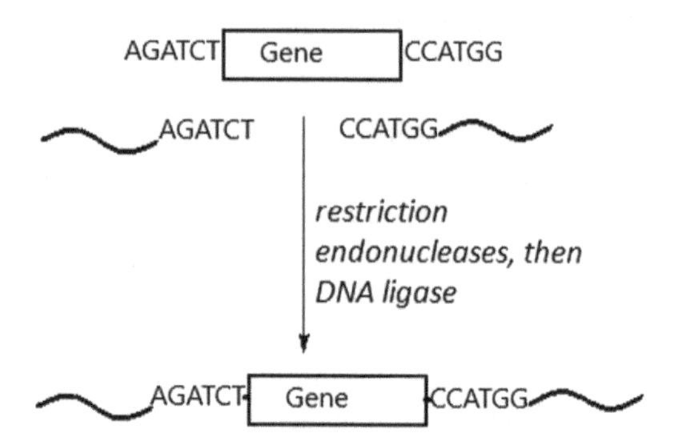

FIGURE 17.3 A gene which is flanked by the same recognition sequences as those found in a plasmid will be inserted into the plasmid at the positions of the recognition sequences, and any nicks in the DNA will be repaired by DNA ligase.

Using restriction sites with this goal requires the researcher to know the positions and types of the sites on the plasmid. Otherwise, it is not possible to guarantee the orientation of the gene as it goes in and out of a plasmid. You would need to know this if you were using the plasmid for different purposes.

PLASMID TYPES AND PURPOSES

Some plasmids serve better functions than others. Plasmids can be used to express proteins, as described. An inducible plasmid in the right cell line can cause up to 10% of the cell mass to be the protein of interest. However, sometimes expression plasmids are not good for high fidelity gene storage or DNA mutagenesis. They do not always produce high copy number, which makes it harder to get material for research work. If you are going to alter a protein by altering the gene, you need to have enough of the gene to manipulate. The expression plasmid might well never be present in large enough abundance to allow you to do this kind of work. Instead, the gene could be in a different plasmid which innately produces high copy number, be mutagenized there, then transferred back into an expression plasmid using the restriction endonuclease technology we have described. Furthermore, some expression plasmids work better for certain proteins than others. There are proteins that, if their expression is more rapid than their ability to assemble into oligomers, will fall out of solution as inclusion bodies. These may comprise 10% of the cell mass, but it all is unusable protein. Such proteins would have been better expressed in a slower expression system which was given longer to operate. This also would have required transfer to a new plasmid. It is even possible that a gene from one species will not fold in another. The gene for human hemoglobin cannot be expressed off of a plasmid in *E. coli* into functional protein, because it will not fold in the proper way in that organism. However, the DNA for hemoglobin can be easily manipulated in *E. coli*, and the gene cut out of a plasmid and inserted into a yeast artificial chromosome (YAC) for expression in a eukaryotic system. This method does result in functional hemoglobin protein, which has been manipulated across species in a prokaryotic system. All these approaches require the use of restriction DNA technology.

RESTRICTION DIGESTS

Mapping a plasmid out involves finding the relative position of restriction sites in the protein. The principle of the technique is that the plasmid can be exposed to a combination of restriction enzymes, and the fragments run out on a DNA agarose gel in order to determine their sizes. If a fragment exists when only a single restriction enzyme is used, but is cut when two restriction enzymes are used, then it can be deduced that the restriction site for the second enzyme lies in between restriction sites for the first one. An example of a gel that has separated the fragments of a restriction digest is shown in Figure 17.4.

In this image, the uncut plasmid runs next to three specimens where the plasmid was cut with single restriction enzymes PstI, EcoRI, and NdeI. It appears to run faster, as if it were smaller, but this is an artifact of the uncut plasmid being circular, whereas the three restriction enzymes each cut at one site, each leading to a single piece of DNA of the same size. When PstI and EcoRI are both used, as seen in the sixth column, the single piece is cut into two, one large and one small. This indicates that the two restriction sites are fairly close to each other, and that there is only one specimen of each site. In the seventh column, PstI and NdeI are both used, and the two resulting pieces of DNA are closer in size, indicating that the restriction sites are further from each other. In the last column, since there do not appear to be two different sized bands of DNA, but what is observed is smaller than the bands for the DNA exposed to only one restriction enzyme, then the two restriction sites must be on nearly opposite sides of the plasmid.

It is the use of logic and mapping out what combinations of observed sizes for the DNA fragments that allow researchers to map out a plasmid. If enough combinations of restriction enzymes are used, a very accurate map of the plasmid can be constructed, even without having to sequence the DNA itself.

Goal

The purpose of this experiment is to map the relative locations of a small subset of the restriction sites on a plasmid that is commonly used for molecular biology work with *E. coli*.

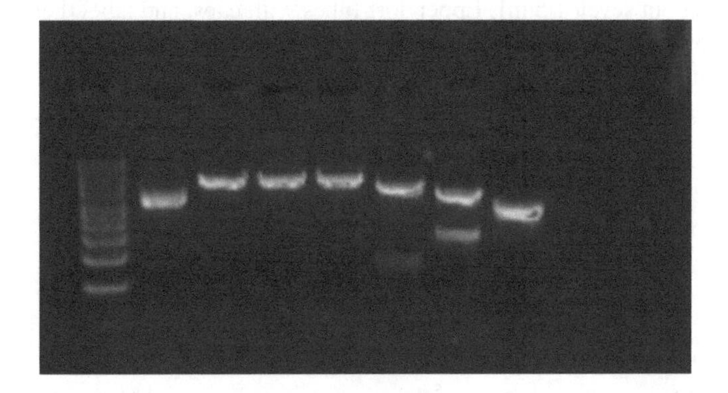

FIGURE 17.4 An agarose gel separating fragments of a plasmid which has been cut with different restriction endonucleases.

Procedure

A. Prepare a 1% agarose gel in 1X TAE buffer. To accomplish this task, you must do the following:
   1. Obtain a mould for an agarose minigel, and a 10-well comb. Determine the volume your mold will hold. Most minigel molds hold between 60 to 100 mL of material. If necessary, use masking tape to create a leak-proof barrier at the ends of the mold, though some types of molds will have rubber gaskets and need no further effort to be leakproof. Be sure the comb is in the position where you want to create wells.
   2. In a 250 Erlenmeyer flask, place 1 g of agarose. Add 2.5 mL of 40x TAE buffer and 97.5 mL of distilled water. Also add 10 μL of 10,000x SYBR SAFE DNA stain.* Swirl the flask to disperse the agarose. It will not dissolve.
      **Other DNA stains work as well, but most are highly carcinogenic. Ethidium bromide, for example, gives extremely intense staining results, but is highly hazardous to use and has complicated disposal requirements.*
   3. Microwave the agarose mixture on the highest power setting until the mixture boils. This usually takes less than 1 minute.
   4. Swirl the mixture again. The agarose will dissolve into a molten mixture.
   5. While it is still molten, pour the agarose into the mold to a depth of 1 cm or else to the top of the teeth of the comb, whichever is less.
   6. If there are any bubbles on top of the molten gel, attempt to burst them or sweep them to the side. It is not recommended to use flame in order to burst the bubbles, but many gel crafters do so, anyways.
   7. Allow the gel to cool and solidify. This takes at least 30 minutes.

B. Prepare a digest of your plasmid.
   1. Suspend lyophilized pBR322 plasmid in *sterile, nuclease-free* deionized water, such that it has a concentration of 0.2 mg/mL. This is your DNA stock.
   2. Set a heating block or water bath to 37°C, for use in later steps.
   3. Set out seven 1.5 mL Eppendorf tubes with caps, and label them as follows: control, P, E, N, P+E, P+N, E+N
   4. Into the seven tubes pipet the following amounts of materials:

Tube	Pst 1 Buffer	EcoR 1 Buffer	Nde 1 Buffer	Pst 1	EcoR 1	Nde	DNA	H_2O
Control	–	–	–	–	–	–	8 μL	12 μL
P	5 μL	–	–	1 μL	–	–	8 μL	6 μL
E	–	5 μL	–	–	1 μL	–	8 μL	6 μL
N	–	–	5 μL	–	–	1 μL	8 μL	6 μL
P+E	5 μL	5 μL	–	1 μL	1 μL	–	8 μL	–
P+N	5 μL	–	5 μL	1 μL	–	1 μL	8 μL	–
E+N	–	5 μL	5 μL	–	1 μL	1 μL	8 μL	–

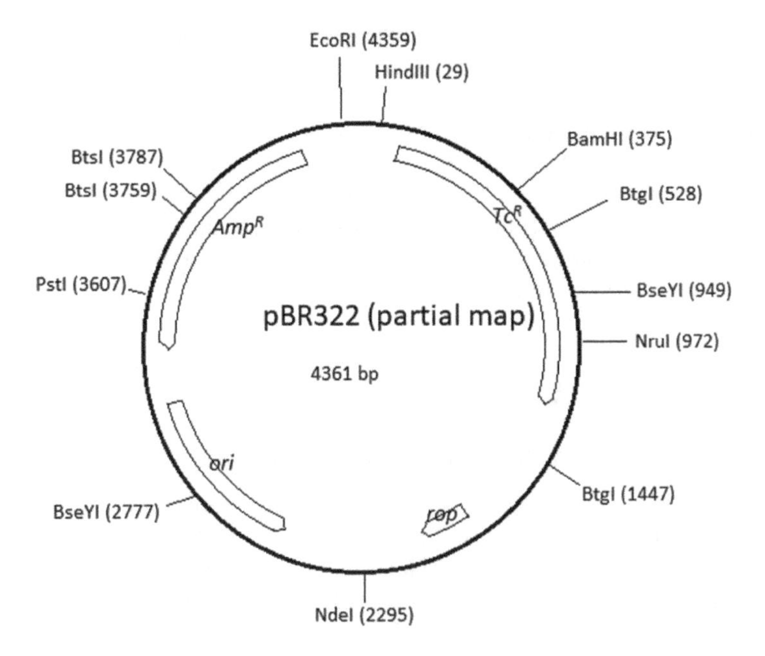

FIGURE 17.5 A partial restriction map of the plasmid pBR322.

Also make a tube with 20 μL of water in it, to serve as a counterweight in later steps.

5. Close the cap and gently shake the tubes in order to mix the component.
6. Place the tubes into a microfuge, balanced against each other. You will use the counterweight in order to do this. Pulse the microfuge at 10,000 rpm for 5 to 10 seconds, in order to push all the liquid to the bottom of the tube.
7. Place the tubes in a heating block or water bath at 37°C for 30 minutes.
8. Remove the tubes from heat.

C. Gel Electrophoresis.
   1. Add 4 μL of loading dye to each tube.
   2. Obtain a DNA ladder. One that ranges from 250 bp to 5000 bp will work best for pBR322, which is 4361 bp in size. If it is not mixed with dye, add 2 μL for each 10 μL of volume.
   3. Place the agarose gel into an electrophoresis device such that the cathode is closest to the well. Current flows from the cathode to the anode, so it will cause the DNA to go through the gel, away from the cathode.
   4. Add enough 1× TAE buffer to cover the gel.
   5. Load 10 μL of the ladder into the well furthest to the left. Then load the wells from left to right in the order: control, P, E, N, P+E, P+N, E+N. The last two wells on the right can remain empty.
   6. Turn on the current and run the electrophoresis device at constant voltage of 100 V. It will take approximately 30 minutes for the dye to reach the bottom of the gel.

7. Remove the gel and image it with a gel imaging device. The SYBR Safe dye allows for imaging by CCD camera when stimulated at 280 nm or 502 nm.
8. Use the DNA ladder to determine the size of the fragments of DNA in each sample. Determine the number and size of each fragment.
9. Use the number and size of the fragments to determine the relative location of restriction sites. Construct a restriction map of pBR322. Compare your map to the known restriction map of pBR322, shown in Figure 17.5 (www.neb.com).

PRE-LAB QUESTIONS

Tutorial: How to Construct a Restriction Map, Logically

Making a restriction map is a game of logically putting pieces together. It is solving a puzzle whose parts are fragments of DNA. The trick is to determine where any site falls between two other sites, and how far away from the edge it is. When you find the order of the pieces, you put them into a circular form for a plasmid or a linear form for linear DNA. In my teaching example I will show you how to do it with a plasmid.

So, suppose you get a plasmid, and you don't know its restriction map. You digest it in separate combinations of the restriction endonucleases XhoI, BamHI, and SfoV. You run them out on a gel with a molecular weight ladder that tells you the size of the pieces, and get the following results:

Plasmid only	1 piece	10 kbp
With XhoI	1 piece	10 kbp
With BamHI	1 piece	10 kbp
With SfoV	2 pieces	5.5 kbp and 4.5 kbp
With XhoI and BamHI	2 pieces	7.5 kbp and 2.5 kbp
With XhoI and SfoV	3 pieces	4.5 kbp, 4 kbp and 1.5 kbp
With BamHI and SfoV	3 pieces	5.5 kbp, 3.5 kbp and 1 kbp
With XhoI, BamHI and SfoV	4 pieces	4 kbp, 3.5 kbp, 1.5 kbp and 1 kbp

Each lane lets us interpret some results. As we get more results, we can interpret more.

Plasmid only	The plasmid total is 10kbp in size. Everything should add to this.
With XhoI	0 or 1 XhoI site. It is the right total either way.
With BamHI	0 or 1 XhoI site. It is the right total either way.
With SfoV	2 cut sites. One is 4.5 kbp from the other one in one direction 5.5 kbp in the other direction. At this point we do not know the positions of the SfoV sites relative to the other sites, if those even exist.

(*Continued*)

With XhoI and BamHI	This proves that there is one XhoI site and one BamHI site. Look at the one piece in the XhoI which was 10 kbp. It is gone, and there are two pieces. The Xho1 site is 2.5 kbp from the BamHI site in one direction and 7.5 kbp in the other direction. The total size of the two pieces is 10 kbp.
With XhoI and SfoV	This one shows the relative position of the XhoI site to the SfoV sites. It is located on the SfoV piece that is 5.5 kbp long. That piece was in the SfoV digest, but not in this one. A missing piece always tells you where the new restriction site is located. What is more, the XhoI piece is 4 kbp from one end of the 5.5 kbp piece, and 1.5 kbp from the other end. We do not yet know which end it is at, though. There are still two possibilities.
With BamHI and SfoV	This one shows the relative position of the BamHI site to the SfoV sites. It is located on the SfoV piece that is 4.5 kbp long, *i.e.* the OTHER piece than the one XhoI is located on. It is 1 kbp from one end and 3.5 kbp from the other end. Unlike the previous logic round, when we did not know which of two possibilities were true for where the BamHI site is located, this time we DO know where it is. You see, we know that the BamHI site is 2.5 kbp from the XhoI site. BamHI is 1 kbp from one SfoV site, and proceeding through that SfoV site, we go a mere 1.5 kbp further until we reach the XhoI site, which you will notice is consistent with where we knew the XhoI site had to be.
With XhoI, BamHI and SfoV	Actually, we had the restriction map in the last round. If you put the XhoI site at the top of a circle that is 10 units in circumference, you would go 1.5 kbp until you reach a SfoV site, then go a further 1 kbp until you reach a HindIII site, then go a further 3.5 kbp until you reach the next SfoV site, then go 4 kbp further until you reach your original XhoI site. And behold, these are the sizes of the fragments in this column, confirming our answer, as seen in Figure 17.6.

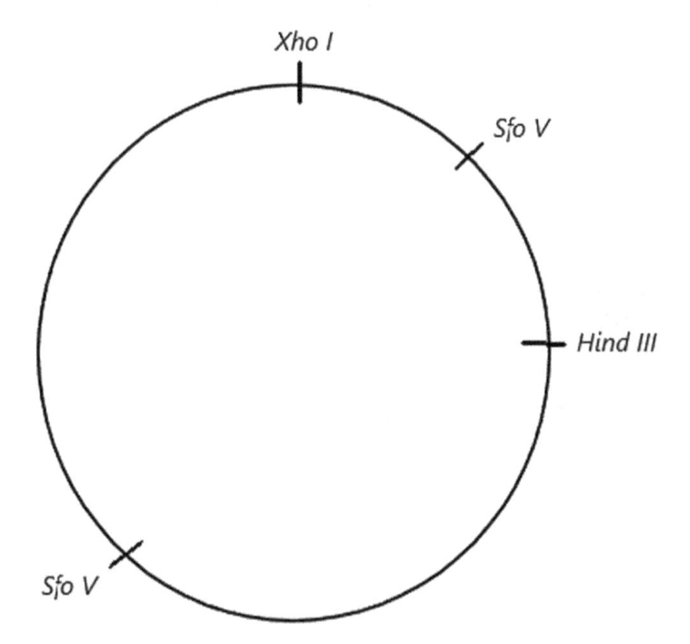

FIGURE 17.6 The solution to the tutorial is a plasmid which should appear as shown in this figure.

Do the following problem before attempting this lab:

1. You obtain a plasmid from some source, and you know NOTHING about it. You digest it with varying combinations of HindIII, EcoRI, and PstI and run the products out on a 0.7% agarose gel stained with SYBR safe stain, with a 1kbp ladder in the leftmost lane. You get the following results:

Plasmid only:	1 band	12 kbp
+HindIII:	2 bands	8 kbp and 4 kbp
+EcoRI	1 band	12 kbp
+PstI	2 bands	11.5 kbp and 0.5 kbp
+HindIII/EcoRI	3 bands	8 kbp, 2.5 kbp, and 1.5 kbp
+HindIII/PstI	4 bands	4.5 kbp, 4 kbp, 3 kbp, and 0.5kbp
+EcoRI/PstI	3 bands	6 kbp, 5.5 kbp, and 0.5 kbp
+HindIII/EcoRI/PstI	5 bands	4.5 kbp, 3 kbp, 2.5 kbp, 1.5 kbp, and 0.5 kbp

a. Draw the gel making the measurements as accurately as possible:

b. State the number of restriction sites:
   i. HindIII
   ii. EcoRI
   iii. PstI

c. Identify the palindromic sequences of:
   i. HindIII
   ii. EcoRI
   iii. PstI
   iv. NdeI

d. Draw the restriction map of the plasmid. (*i.e.* make a circle on which you draw the restriction sites and indicate the distance between sites; by convention they are drawn clockwise.)

18 Western Blotting

In a previous chapter, the technique known as "ELISA" was described. This is an immunoabsorption assay that is good for revealing the presence of any protein, whether it has an enzymatic activity or not. As was mentioned in that chapter, the power of the ELISA technique comes through most powerfully when in combination with other techniques, of which one will be studied in this laboratory: western blotting.

In another previous chapter, we discussed the use of polyacrylamide gels, and examined the mobility of proteins in SDS-PAGE analysis. Normally, when proteins are migrating across a polyacrylamide gel, they are repelled from the anode in proportion to their charge, but their movement is also in proportion to their mass. Thus, the speed of migration is proportional to the ratio of charge to mass, and not just the mass. When sodium dodecyl sulfate (SDS) is added, it adds so much negative charge to the protein that all proteins have the same charge. The only thing controlling their rate of migration, therefore, is their mass. This makes identification of the bands considerably easier than it would be in a native gel, which does not have SDS.

The disadvantage of an SDS-PAGE gel is that the sodium dodecyl sulfate also disrupts the structure of the protein. A native gel, on the other hand, preserves the structure and therefore the activity of the target proteins. If the protein is in its native conformation, then an antibody which binds to a surface epitope of that protein will be able to recognize and bind it after it is separated. The same antibody could not have been used if the protein were complexed to SDS, unless you are fortunate enough to have a completely linear epitope. This is not always the case, and if so, you will have to separate the proteins using native gels before you can use antibodies to image them

The best combined use of gel electrophoresis and immunostaining for imaging purposes is the technique known as "western blotting". This name was given in a somewhat tongue-in-cheek fashion, because it sounds like it refers to a point on the compass. This is not the case. A technique for identifying specific DNA sequences that also used electrophoresis was developed by Edwin Southern (1975) and became known as the "Southern blot." It is the only one of the "compass point" blots that is named after a person. In 1977, a somewhat similar combination was developed to identify RNA sequences, and was dubbed the "northern blot" (Alwine et al., 1977). The technique now known as the "western blot" was developed in 1979 in order to detect proteins (Towbin et al., 1979). Strong resistance kept any technique from gaining the term "eastern blot" for many years, but by 2001 it had been adapted for a technique that imaged post-translational modification of proteins (Shan et al., 2001) by a somewhat large community. The names have nothing to do with where the techniques originate.

In western blotting, as with all other forms of blotting, there is an initial electrophoresis technique, followed by steps that allow imaging of the products in a specific way. The electrophoresis step of western blotting may be SDS-PAGE, native PAGE, or even the technique known as "isoelectric focusing". In this latter technique, the gel has a pH gradient, and as the protein passes through, it comes to a point where the pH equals the isoelectric point. At that position, the protein has no net charge and so does not move further. Whatever technique of electrophoresis that is used, it is important that the proteins be separated from one another in a fashion that allows the epitope to remain intact. Otherwise, the later steps will not work.

Having separated the proteins electrophoretically, the proteins must be removed from the gel onto a membrane. The membrane historically was nitrocellulose paper, but polyvinylidene difluoride (PVDF) is the more modern choice, and offers many advantages. This is a necessary step because the antibodies would not rapidly be able to penetrate the matrix of the polyacrylamide gel and bind to the target protein. It is also the part of the technique known as "blotting", from which the rest of the name is derived. The gel is sandwiched with the membrane between two electrode plates, and voltage is applied perpendicularly to the gel, such that the proteins move out of the gel and directly onto the membrane, rather than further up and down the gel. Then the membrane is saturated by some non-specific protein that will keep antibodies from accidentally binding to any random position. This saturation step, known as "blocking", is usually done by briefly soaking the membrane in a solution of casein, a group of proteins obtained from milk. The apparatus is diagrammed in Figure 18.1.

At this point, antibodies can be added for the target protein of interest. As described above, the primary antibody is first added, by letting the blocked membrane soak briefly in a solution containing the antibody. A washing step removes any antibody that did not bind to the target, and then the membrane is exposed to a solution containing the secondary antibody, also as described. The membrane is washed again, so that only the secondary antibody that has bound to the primary antibody will remain. Finally, the imaging technique of choice is applied, either the induction of fluorescence, or detection of radioactivity, or chemical imaging in the case of horseradish peroxidase or other similar imaging enzymes. The imaged spots that appear will mark the position of the protein of interest.

In this laboratory, students will use western blotting to locate and identify the protein "superoxide dismutase 1" from a bovine source. It will be given as a blind sample along with two other proteins which should not react with the antibodies. Having identified which of the samples contains the SOD1 protein, students will see if they can identify it in a homogenate from living tissue (Figure 18.1).

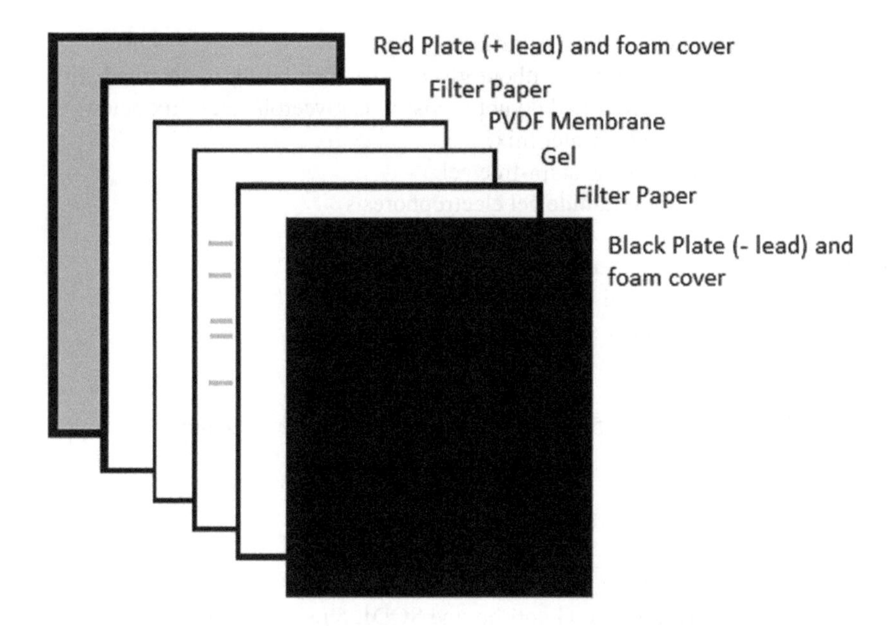

FIGURE 18.1 The design of the stacked layers used to blot proteins onto a membrane.

GOAL

The purpose of this experiment is to determine the presence of the enzyme superoxide dismutase 1 (SOD1) on a native gel.

Set-Up

The instructor will have to prepare the unknown samples and the antibodies immediately before the start of the lab period.

Materials

Part 1

Unstained protein standards for native gel, MW range from 20 kDa to 1028 kDa
An unknown which is 0.2 μg/mL SOD1 in 0.02 M phosphate buffer, pH 7.4
An unknown which is 0.2 μg/mL bovine serum albumin in 0.02 M phosphate buffer, pH 7.4
An unknown which is 0.2 μg/mL L-lactate dehydrogenase in 0.02 M phosphate buffer, pH 7.4
1 gram of bovine liver homogenized in 100 mL of 0.02 M phosphate buffer, pH 7.4

Native Gel Tris-Acetate-EDTA buffer, pH 8.5
4x native gel loading buffer without glycerol (if such buffer is obtained with glycerol, use 2x buffer and do not prepare the glycerol/water mix below)
25 mL of 50:50 glycerol/water mix.
10% polyacrylamide TGX stain-free gel
Apparatus for polyacrylamide gel electrophoresis

Part 2

Apparatus for horizontal blotting gel electrophoresis
PVDF membrane paper
Whatman 3 MM paper
Methanol
0.02 M Tris buffer pH 8.5 with 0.15 M glycine and 20% methanol

Part 3

10% Tween 20 in deionized water
casein
0.02 M phosphate buffer, pH 7.4
Primary antibody, rabbit IgG anti-bovine SOD1, 5 μg/mL in 100 mM carbonate/bicarbonate buffer, pH 9.6
Secondary antibody, goat anti-rabbit antibody, conjugated to horseradish peroxidase, 5 μg/mL in 100 mM carbonate/bicarbonate buffer, pH 9.6
2,2′,5,5′-tetramethyl benzidine (TMB), 5 μM in 0.02 M phosphate buffer, pH 7.4
2 M H_2SO_4 (optional)

Instructions

Part 1: Native Gel Electrophoresis

1. Construct the apparatus for running electrophoresis on a polyacrylamide gel. Set up a polyacrylamide gel as has been described in a previous chapter ("Polyacrylamide Gels") using native gel buffer instead of SDS gel buffer.
2. Prepare your samples by pipetting into a 1.5 mL Eppendorf tube with a cap the following ingredients: 10 μL of a sample, 5 μL of 4x loading dye, and 5 μL of glycerol mix (or 10 μL of a sample and 10 μL of 2x loading dye with glycerol). Prepare such a sample for each unknown and for the liver homogenate. Mix and centrifuge at 10,000 rpm for 10 seconds to collect the samples at the bottom of the tubes.
3. Set up the get by pipetting 10 μL of the MW markers in the well at the left, then 10 μL of each of the unknowns into the next three wells, and 10 μL of the liver homogenate into the next well. Leave the remaining wells empty.
4. Run the electrophoresis protocol to separate the proteins, as has been described. Detach the gel from the glass or plastic plate in which it was cast, and keep it damp but not submerged while moving on to Part 2. Alternatively, the gel and its casting plates may be wrapped in buffer-soaked paper towels and refrigerated in a plastic bag until a later lab period for use in Part 2.

5. If you did detach the gel at this time, image the gel using a stain-free gel imager. If you are saving your gel, then do this step at the start of Part 2 in a later lab period, but you will sacrifice some resolution if you do it this way.

Part 2: Blotting

6. Cut two sheets of Whatman 3 MM paper so that they are the same size as the gel.
7. Cut a sheet of PVDF paper so that it is also the same size. Soak the PVDF in methanol for 2 minutes. Try to use tweezers to manipulate the PVDF paper at all points, so that proteins in your skin do not adhere to it.
8. Pour Tris/Glycine/Methanol buffer into a shallow dish large enough to fit the PVDF membrane. Transfer the membrane to this buffer and soak it for 3 minutes.
9. Assemble the stack of transfer materials by laying down the following layers, in order:
   a. The positive (red lead) plate of the transfer electrophoresis device
   b. One sheet of Whatman 3 MM paper
   c. The gel
   d. The PVDF membrane
   e. The second sheet of Whatman 3 MM paper
   f. The negative (black lead) plate of the transfer electrophoresis device (At each stage as you lay down an item, remove any air bubbles that lie between the layers. This may be done by rolling a smooth pen, pencil, or test tube, or by gentle massaging. If you do not, you will not get an even transfer.)
10. Check to make sure that the Whatman 3 MM paper on one end of the stack does not in any way contact the other side, so that you do not create a short circuit.
11. Assemble supports and sandwich clamps to hold the gel in place in the horizontal electrophorimeter.
12. Add transfer buffer to submerge the electrophoresis cell. If you can, make sure this buffer is cold in order to increase the efficiency of transfer.
13. Apply potential of 100 V for one hour. With a minigel, this should result in transfer to the membrane. If you go longer, you run the risk of pushing your proteins off the PVDF paper.
14. Remove the membrane from the stack and save it for the next step. Alternatively, it may be wrapped in buffer-soaked paper towels and refrigerated in a plastic bag until another lab period for use in Part 3. A better stopping point, however, occurs in the middle of Part 3.

Part 3: Immunostaining

15. Prepare a washing buffer by combining 500 mL of 0.02 M phosphate buffer, pH 7.4 with 0.5 mL of 10% Tween 20.
16. Use the buffer you just prepared to create a blocking buffer by adding 5 g of casein to 95 mL of the buffer, and stir to suspend the casein.
17. Create a primary antibody solution by combining the following combination: 90 mL of washing buffer, 10 mL of blocking buffer, and 0.1 mL of the primary antibody.

18. Place the PVDF membrane into a shallow dish and cover it with washing buffer. Agitate it for 5 minutes to wash the membrane, then remove it from the washing buffer. Discard the buffer.
19. Place the PVDF membrane back into the shallow dish and cover it with blocking buffer. Incubate it at 4°C for 1 hour with agitation, such as a rocking plate. Remove it from the blocking buffer and rinse one time with washing buffer for only 5 seconds.
20. Place the PVDF membrane back in the shallow dish and cover it with the primary antibody solution. Incubate it at 4°C overnight for high quality results. Do not incubate longer than 18 hours
21. Create a secondary antibody solution by combining the following combination: 100 mL of washing buffer, and 0.05 mL of the secondary antibody.
22. Remove the PVDF membrane from the primary antibody solution and wash with wash buffer, as in Step 19, for only 5 seconds. Repeat this wash five times.
23. Place the PVDF membrane back in the shallow dish and cover it with the secondary antibody solution. Incubate it at room temperature for 1 to 2 hours, on a rocker plate to provide agitation.
24. Remove the PVDF membrane from the secondary antibody solution. Wash the membrane with washing buffer five times, for 5 seconds each time, as before.
25. Place the PVDF membrane in the shallow dish a final time, and cover it with TMB solution. Incubate at room temperature with agitation until the color appears strongly.
26. Remove the membrane from the TMB solution and wash it with running water.
27. Image the membrane in order to quantify the location of the SOD1 in the unknown samples. Compare this image to the image of the gel taken by stain-free imaging at the end of Part 1. Determine if SOD1 is also present in the beef liver lysate. Use the standards to identify which unknown was albumin and which was L-lactate dehydrogenase. Determine if they are in the liver lysate as well.

REFERENCES

Alwine, J. C., Kemp, D. J., & Stark, G. R. Method for detection of specific RNAs in agarose gels by transfer to diazobenzyloxymethyl-paper and hybridization with DNA probes. *Proceedings of the National Academy of Sciences*, *74*(12), (1977): 5350–5354.

Shan, S., Tanaka, H., & Shoyama, Y. Enzyme-linked immunosorbent assay for glycyrrhizin using anti-glycyrrhizin monoclonal antibody and an eastern blotting technique for glucuronides of glycyrrhetic acid. *Analytical Chemistry*, *73*(24), (2001): 5784–5790.

Southern, E. M. Detection of specific sequences among DNA fragments separated by gel electrophoresis. *Journal of Molecular Biology*, *98*(3), (1975): 503–517.

Towbin, H., Staehelin, T., & Gordon, J. Electrophoretic transfer of proteins from polyacrylamide gels to nitrocellulose sheets: Procedure and some applications. *Proceedings of the National Academy of Sciences*, *76*(9), (1979): 4350–4354.

Index